About Island Press

Island Press is the only nonprofit organization in the United States whose principal purpose is the publication of books on environmental issues and natural resource management. We provide solutions-oriented information to professionals, public officials, business and community leaders, and concerned citizens who are shaping responses to environmental problems.

In 1994, Island Press celebrates its tenth anniversary as the leading provider of timely and practical books that take a multidisciplinary approach to critical environmental concerns. Our growing list of titles reflects our commitment to bringing the best of an expanding body of literature to the environmental community throughout North America and the world.

Support for Island Press is provided by The Geraldine R. Dodge Foundation, The Energy Foundation, The Ford Foundation, The George Gund Foundation, William and Flora Hewlett Foundation, The James Irvine Foundation, The John D. and Catherine T. MacArthur Foundation, The Andrew W. Mellon Foundation, The Joyce Mertz-Gilmore Foundation, The New-Land Foundation, The Pew Charitable Trusts, The Rockefeller Brothers Fund, The Tides Foundation, Turner Foundation, Inc., The Rockefeller Philanthropic Collaborative, Inc., and individual donors.

Wild Forests

Wild Forests

CONSERVATION BIOLOGY
AND PUBLIC POLICY

William S. Alverson

Walter Kuhlmann

Donald M. Waller

Foreword by
Jared Diamond

ISLAND PRESS WASHINGTON, D.C. / COVELO, CALIFORNIA

Alverson, William Surprison.
    Wild Forests conservation biology and public policy/ William S. Alverson, Walter Kuhlmann, Donald M. Waller; foreword by Jared Diamond.
        p.   cm.
    Includes bibliographical references (p.   ) and index.
    ISBN 1-55963-187-2 (cloth: acid-free paper). —
ISBN 1-55963-188-0 (paper: acid-free paper)
    1. Forest conservation—United States.   2. Forest ecology—United States.   3. Forest management—United States.   4. Biological diversity conservation—United States.   5. Forest policy—United States.   6. United States. Forest Service.   7. Forest reserves—United States—Management.   I. Kuhlmann, Walter.   II. Waller, Donald M.   III. Title.
SD412.A38   1994   94-8950
333.75'16'0973—dc20   CIP

Manufactured in the United States of America
10   9   8   7   6   5   4   3   2   1

To Kenneth Bowling and Harriet Irwin,
teachers extraordinary.
—W.S.A.

For Debby, Hannah, and Charlie,
who see the forests with me.
—W.K.

To Henry S. Horn.
—D.M.W.

Contents

Figures and Tables

Figures

Tables

Foreword

One day many years ago, I had a memorable discussion in a government office in the capital of a developing Pacific nation, the Solomon Islands. My conversation partner was the government's chief botanist, and our subject of conversation was forestry, the industry responsible for producing the largest share of the country's foreign exchange earnings.

My friend explained to me his problems with the country's most valuable timber tree, the species known as *Campnosperma brevipetiolata*. His forestry department had tried to grow the tree from seeds, but the seedlings never germinated. As a result, Solomon forestry methods still depended on cutting the tree in rainforests where it grows mixed with many other tree species.

How, I asked my friend, did those recalcitrant seeds manage to propagate themselves abundantly in nature? Did my friend ever see seedlings in the forest? Of course he did, usually in clumps. When he looked up, he often noticed that a branch above the clumps of seedlings was stained white with excreta, evidently from some bird or bat that habitually roosted there. Thus *Campnosperma brevipetiolata* must be one of the many tree species whose seeds are normally dispersed by animals and will germinate only if first passed through an animal's gut.

It was a striking example of our ignorance of some economically important biological facts. Here we were talking about a tree that was the major source of income for a cash-starved country. Yet scientists lacked the biological knowledge necessary to grow the tree.

That conversation took place 18 years ago, in 1976. I wish I could report that in the intervening years Solomon foresters have learned to grow *Campnosperma brevipetiolata* and have planted self-sustaining plantations of it. Instead, the Solomon government has postponed the day of reckoning and embarked on a crash course of cutting its natural forests. The forests are being treated as a nonreplenishable mine, rather than as a renewable resource. By the end of the next decade, the Solomons' accessible forests will be gone. The Solomon government will then be back to its 1976 problem of learning how to grow *Campnosperma brevipetiolata* —but under much less favorable circumstances. Most of the natural stands and their genetic diversity will have been eliminated. Thanks to the human population boom, the Solomon Islands will then have far more people to feed and fewer sustainable sources of revenue with which to feed them.

Our first temptation may be to dismiss this public policy decision as something that could happen only in a remote, exotic, foolish, third-world country, where there are few scientists and where government leaders are too ignorant or short-sighted to recognize their national self-interest. The United States is better off—but only slightly, and not better off enough.

America has had its own history of rapacious, non-self-sustaining forestry. It is not something of the past. Our remaining native forests continue to be logged as if they were mines. We still lack a national biological inventory, and we are only just beginning efforts to develop one. We don't even know enough to manage our surviving forests. The oldest trees in our remaining small tracts of mature eastern forest—Heart's Content Forest in Pennsylvania (see Chapter 3), Fontenelle Forest in Nebraska, Hutcheson Memorial Forest in New Jersey, and others—are being replaced by different tree species, for reasons that remain controversial. Fire control, high deer numbers, and elimination of top predators all undoubtedly contribute, but the relative roles of these factors are uncertain.

While we don't know enough to manage our forests, we already do know enough to appreciate that not having to manage them is in the long run the cheapest, most successful strategy. That requires forest tracts big enough to manage themselves. When we stamp a cookie-cutter on a forested landscape, clearcut outside the cookie edge, and hope that the forest cookie will remain pristine, we are guaranteed disappointment. Small forest tracts *cannot* sustain themselves in a pristine condition because they were originally being sustained by processes beyond their boundaries. We eliminated those processes when we isolated the tracts. The processes include inputs of nutrients—maintenance of populations of top predators that live at low numbers but that disproportionately affect community structure—and the patch dynamics that maintain large landscapes in a steady-state mosaic.

Even the ecosystems of our biggest national parks, such as Yellowstone, may not be self-sustaining. They are doubtfully big enough to maintain unmanaged populations of grizzly bears and wolves, which affect large herbivore numbers, which in turn affect the vegetation on which other species depend. But our well-publicized difficulties in managing Yellowstone fade into insignificance when compared with our difficulties in managing vest-pocket scraps of natural forest.

Thus, we face a dilemma. A hands-off policy won't work, yet we don't know enough to formulate a successful hands-on policy. We can at least minimize that dilemma in large tracts and landscapes.

I found this book to be a gripping case study of that dilemma. In the pages ahead, you will find all the elements of a thriller novel. Powerful, well-intentioned people, with diametrically opposed but strongly held views, are in conflict about what to do about our wild forests. The conflict involves big legal, political, economic, and ideological issues. It also involves big issues at the cutting edge of scientific knowledge. The ultimate test of our scientific knowledge of forests will be whether we can manage them or successfully avoid having to manage them. Can developed countries set an example for developing countries, such as the Solomon Islands? Or, are we all condemned to perpetuate the same follies?

Jared Diamond

Preface

This book was born out of our shared concern for protecting biological diversity. Our work and wanderings in the forests of northern Wisconsin, Minnesota, and Michigan have given us a deep regard for the Great Lakes Northwoods. Although these lands are diminished from their presettlement biotic richness and are not nearly so nationally prominent as the old-growth rainforests in the Northwest, they remain worthy of our best efforts to return major portions to old-growth conditions.

We share the conviction of most contemporary ecologists that the biggest threats to diversity currently lie in the tropics or, on this continent, in the ancient forests of the Pacific Northwest. Nevertheless, we believe the biological and public policy issues we face in the northern Great Lakes forests justify the examination given in this book.

Despite being virtually completely logged on one or more occasions, our northern forests harbor a diverse and unique biota, including sensitive species and ecological processes that warrant conservation efforts. Secondly, future work of conservation biologists increasingly will consist of restoring impaired or fragmented habitats in already disturbed areas. In our northern forests, we now realize that maintaining the valuable elements of current diversity and restoring old-growth conditions in the future are congruent endeavors. Thus, in these forests we encounter the problems of appreciating the complexity of forest ecosystems, articulating the management principles that will restore (and thus maintain) ecological characteristics, and incorporating these ideas into laws, regulations, and other forms of public policy—problems common to all corners of the globe.

Our baptism into the fractious arena of how conservation science should affect land management policies came in 1985 upon reading draft versions of long-term management plans and associated Environmental Impact Statements for Wisconsin's two National Forests. We were disappointed that management proposals by the U.S. Forest Service showed superficial concern for biodiversity issues and failed adequately to incorporate what we considered mainstream ideas from ecological and conservation science.

Our disappointment and concern grew when, in seeking to influence forest policy, we met obstinate resistance to what we had assumed were accepted scientific concepts and a fair reading of current laws and regulations. A long, and still pending, struggle ensued over how the Agency (i.e., the Forest Service) should respond to biological diversity issues on the two

Wisconsin National Forests and, by extension, over the 77 million hectares contained in the other National Forests (Mlot 1992). The Wisconsin cases demonstrate how this key federal land management agency has responded to the challenges posed by conservation biology, and reveal the difficulties we face in reforming forest management.

At the center of these cases and this book is the question of whether our goal should be the crafting of increasingly better engineering techniques to achieve desired population levels of preferred species (whether preferred for their economic value, beauty, or rarity), or whether we should focus on restoring and maintaining general ecological conditions and thereby allow natural processes to determine and maintain species mix and population levels. This fundamental question is older than Muir, Pinchot, and Leopold, and as current as the pleadings of many of the Forest Service's own best scientists in the northern spotted owl controversy.

As the book's title makes clear, our understanding of the grave consequences of past efforts at forest engineering, our belief in the critical importance of the multitude of unseen species and processes, and our knowledge of how little we really know about forest ecology all lead us to pursue wild (i.e., unengineered) conditions as the surest way to achieve a return to the biological conditions which brought us the species richness we now enjoy.

Our preference for wild forests is the result of a hard-headed scientific assessment of the ecological processes that characterize forests and of the substantial past and likely future losses of biodiversity. After reviewing and participating in many studies of the results of invasive forest management, many field surveys of rare populations, and many meetings with forest managers bent on "doing better next time," we conclude that biodiversity protection in our forests can best be achieved by restoring natural ecological processes over large areas through largely passive methods of management.

Not surprisingly, such an approach is not well received by those in the field who have managed forests for decades in response to the commodity goals society has set for them. Taught in fine educational institutions that forests are far better off as a result of extensive human intervention, many forest managers view conservation biology as a pretext designed to indict past management decisions and ban all logging from public land.

We ask the reader to set aside his or her prior views on what forests are and what is best for their future, remain open to the evidence and ideas presented, and take a short holiday from the contentious and emotional arena of public discourse over forest management. To those who would at least join us in agreeing that protection of biodiversity must be a primary goal of forest management, we suggest that all of us focus on the future, and we welcome your interest in, and critique of, our case for wild forests.

Acknowledgments

We list ourselves alphabetically as authors to indicate our equal contributions to this effort. It is a pleasure to acknowledge the cooperation and contributions of the many other individuals who helped generate this book. Our initial involvement in forest management issues began in 1980 with rare plant surveys in the Chequamegon and Nicolet National Forests conducted by W. Alverson, Mark Jaunzems, Emmet Judziewicz, Robbin Moran, and Stephen Solheim, with funding from the USDA Forest Service. Draft and Final Land and Resource Management Plans appeared for these Forests in 1985 and 1986, precipitating a dialogue with the staffs of the Chequamegon and Nicolet that continues to this day. Despite formal U.S. Forest Service opposition on many issues, the commitment, interest, and open-mindedness of many Forest Service staff members have inspired us and helped to sustain our involvement. We especially appreciate former Chequamegon National Forest Supervisor Jack Wolter's honesty and willingness to consider new approaches.

Our involvement in forest issues was also fueled and sustained by the dedication and comraderie displayed by our partners in the Wisconsin Forest Conservation Task Force, particularly Sharon Clark Gaskill, George Hall, Emmet Judziewicz, Cy Lyle, and Stephen Solheim. This group evolved into a virtual graduate seminar on the application of science to public forest policy and benefitted continually from support provided by the John Muir Chapter of the Sierra Club and the Wisconsin Audubon Council and its many local societies. Stephen Solheim initiated and helped to develop many of the ideas explored in this book. We gratefully acknowledge his many insights about the real causes of phenomena, whether scientific, economic, or cultural, and in particular his perceptions of nature described in Chapter 2. Hugh Iltis inspired our efforts by setting a compelling example of what it means to be committed to conservation, including its political ramifications. Bonnie Wendorff was of considerable help in briefing the Wisconsin forest cases in federal court.

Revising our unruly manuscript was greatly aided by the constructive and comprehensive reviews provided by Ray Guries, Cara Nelson, and an anonymous reviewer. Although we did not always follow their suggestions, their efforts far exceeded what might have been expected from them and have greatly improved the book. Kiva Adler, Caitilyn Allen, Harvey Ballard, Tom Brandner, Dan Cottam, Kristen Hershbell, Dale Kushner, Susan Lohafer, and Robin Phylliky assisted us by reading and commenting on parts of the manuscript. Working well into the wee hours, Tom Brandner graciously re-organized the extensive reference list. Kandis Elliot read sections and con-tributed her artistic skills by drafting several of the figures. Jim Bennett pro-vided several key references. University of Wisconsin law students Alice Thomas, Greg Monday, Peter Ludwig, Karl Mueller, and David Ansel as-sisted with legal research and proposals for avenues of legal reform.

We are particularly grateful to our editor, Joe Ingram, for his enduring confidence in, and encouragement for, our efforts and his patient forbear-ance. Heartfelt thanks are due Caitilyn, Charlie, Debby, Hannah, and Kiva for their support, and tolerance of our many absences, during the genesis of this book.

Finally, we would like to acknowledge the other scientists and conserva-tionists, too numerous to name individually, who shared with us their con-cerns with the thorny practical questions surrounding forest management. While they do not all agree with the ideas presented here, their interest and involvement encouraged us in our efforts and provided a vision for what forest management might become. We particularly commend those coura-geous scientists and foresters within the Forest Service who share our concern for forest biodiversity and put at risk their personal careers to advance the protection of native plants and animals.

Introduction

This book reflects a somewhat unique marriage of science and public policy. It is not a textbook on forest biology, a history of Forest Service policy, or a law review article, but rather something of a synthesis of all three. Too often, those concerned about protecting diversity in our forests, whether as activists or agency personnel, have been schooled in one element of the debate over diversity, leaving them ill-equipped to address the full range of issues relevant to protecting diversity through forest management.

For years, we have heard of the essential need to protect habitat or ecosystems, but these vague entreaties are rarely backed up with the hard facts about the mechanisms by which species and other components of diversity are lost. At the same time, many fine scientists who understand contemporary threats to biodiversity are either unwilling to enter the realm of advocacy or unable to articulate how diversity protection should fit within the history, culture, and law governing how forest managers approach their task. We have attempted a synthesis of science, policy, and law in a long campaign, which surely will continue for many years to come, seeking greater protection for the diversity in the Wisconsin National Forests. It is our view that such a comprehensive approach will be indispensable for future efforts, whether from within land management agencies, by outside activists, or in finding common ground among all parties, if we are to employ the teachings of conservation biology to effect reforms in land management.

Foresters exert great influence over the composition and structure of our forests, especially on public lands where their power and responsibility extend across millions of hectares. Their job, however, has grown more difficult as they attempt to balance multiple and often competing concerns regarding the productivity and sustainability of their forests. Conflicts often exist as foresters seek simultaneously to meet demands for high outputs of timber and to sustain viable populations of all forest creatures. This book examines ways to satisfy these demands once the protection of biodiversity is accepted as a legitimate and important goal.

In Part I we consider past, present, and future aspects of forest management as they relate to conserving biodiversity. Chapter 1 surveys patterns of deforestation in North America over the last three centuries and associated changes in forest management in an attempt to explain how these contributed, directly and indirectly, to conspicuous losses in biological diversity. Our ecosystems have yet to recover fully from these losses and the cascading ecological changes that came in their wake. Nevertheless, they may represent only the tip of an iceberg of ongoing biological losses considered in Chapter 2. Biologists are now discovering that forests suffer many other indirect and inadvertent losses in response to frequent logging. Furthermore, our knowledge of older forests remains too incomplete to fully assess the extent or nature of these losses. In Chapter 3 we explore alternative ways in which the more ecologically sensitive forest managers of tomorrow may seek to protect those elements of diversity threatened by anthropogenic disturbance.

As foresters increasingly concern themselves with the health of forest ecosystems, they confront a central question: When and to what degree do conventional patterns of management threaten biodiversity? To understand when and how management might threaten ecological values, we need to examine the ecological mechanisms that act to sustain, or reduce, the diversity of plant and animal communities. In Part II we review some of what is known regarding how ecological systems gain and lose diversity. Many of these threats revolve around disruption of historical patterns of disturbance (Chapter 4). Other threats derive from the fact that human disturbance and associated edge habitats boost populations of many weedy plants and animals, including edge-loving herbivores, predators, and parasites that threaten the persistence of species usually found in forest-interior habitats (Chapter 5).

In Chapter 6 we explore how populations often suffer from increasing isolation or restricted ranges as their forest habitats become more fragmented. While we concentrate our discussion on biological diversity, we also touch on some implications of ecological findings regarding how silviculture affects soil nutrients, long-term forest productivity, and the net carbon balance of young and old forests. To keep track of the successes and failures in dealing with the processes described in Chapters 4 through 6, we need to improve our methods of assessing the state of forest ecosystems. Forest managers need efficient and accurate means to assess diversity within their forests and to monitor how populations, species, communities, and ecological processes respond to their management practices (Chapter 7).

The extreme forms of "cut and run" logging that dominated in the nineteenth century are widely acknowledged to have been destructive and short-sighted. Nevertheless, there has frequently been the presumption that contemporary "scientific" forestry poses few or no threats to the long-term sustainability and wildlife of a region. We explore these issues in some detail in Part III. In Chapter 8 we review historical developments in forestry and contemporary patterns of silviculture and forest management. In Chapter 9 we analyze the concept of multiple use as defined and practiced by the Forest Service. This dominant paradigm of U.S. forest management is a multifaceted concept, encompassing the philosophical basis of society's relationship to forests (utilization), the status of the land managers (discretion), as well as actual management guidelines (rotation of uses).

Alternative approaches to silviculture and forest management are the topic of Chapter 10. While each of these presents certain advantages, we conclude that the continuing presence of considerable uncertainty (particularly regarding rare and small elements of diversity) dictates choosing management methods that are conservative in the sense of maintaining diversity. In Chapter 11 we present a rationale and criteria for our own prescription: designating large areas as free as possible from major anthropogenic disturbance. It is our opinion that such areas should be maintained until research has amply demonstrated the long-term reliability of other, more manipulative alternatives.

All of these considerations of ecological dynamics, threats to communities, and optimal forest management will be useless if they are not applied. Thus, we consider it essential to devote Part IV to questions of policy and implementation. Chapter 12 explores key aspects of Forest Service policy over the decades and the three principal statutes governing the Agency's management responsibilities in order to determine those elements of history and law that could either aid or impede our application of the conclusions of the preceding chapters. Current laws and codes stipulate that planning and management on these forests should encompass concerns for ecological impacts and biodiversity; yet recent experiences in the cases of Wisconsin's National Forests, described in Chapter 13, demonstrate that the Agency is far from reconciled with its responsibilities to maintain diversity and ambivalent about how science should guide policy.

Our book ends with specific recommendations regarding how our laws and institutions might be better molded to fit our improved understanding of ecological dynamics and accomplish our prescribed means of protecting biodiversity on the ground (Chapter 14). Despite continued opposition in the top echelons of the Agency and some parts of the timber industry, we remain optimistic that public concern for, and understanding of, these fundamental biological issues will eventually translate into effective public policies worthy of the respect of scientists and imitation by other countries around the world. We hope this book contributes in some small way to that progress.

PART I

Whence Biodiversity?

ANY REASONABLE ACCOUNTING of forest biodiversity must take place within an evolutionary framework. Species or local populations of species persist or are lost from forests as a complex function of their evolutionary adaptations and chance. Adaptation and chance occurrence act to determine the fate of biological lineages over long periods of time. The response of each species and population to our actions has been largely predetermined by millenia of genetic adaptation in a pre-human environment. To acknowledge this broad temporal context, we begin our consideration of biodiversity several millenia in the past, when most elements of temperate forest diversity were already in place (Chapter 1). Here we review pre- and post-Columbian losses of diversity, emphasizing the last one to three centuries of European colonization and modification of the landscape.

Given these historical losses, what biodiversity remains in today's forests? In Chapter 2 we review how little we still know regarding most indigenous biological lineages—species and populations. Our ability to protect this native biodiversity will depend not only on our technological cleverness at managing landscapes but also on our awareness of the limitations of the information available to us. In this chapter we point out the immense gaps in our knowledge of forest species, their habitats, and their interactions, and ask what relevance this ignorance may have to present and future forest management.

Having glimpsed something of the losses in diversity suffered by our temperate forests (Chapter 1) and the still largely hidden diversity that remains (Chapter 2), we next consider the fate of diversity in the years to come (Chapter 3). By what means can we maintain and restore forest diversity so as to pass these natural resources on undiminished to our descendants? Thus, we ask in Chapter 3: What is the vision of future forests that we seek?

_____

Forests Then and Now

North American forests have suffered drastic reduction since European settlement. Logging has removed more than 99% of the great virgin forests from the eastern half of the continent and still proceeds apace in the far West. Settlement has stripped most of the lower 48 states of their original large predators and hoofed mammals. How could such drastic losses occur without more public outcry? Why were sentiments for conservation so limited and ineffectual? While answers to these questions fall beyond the scope of this book, it is important to appreciate the nature, extent, and causes of these losses if we are to understand today's continuing threats to biological diversity.

Today, few of us are familiar with the unique sights, sounds, and smells of primary forests. Although many small, remnant stands still appear majestic, they lack a full complement of their original bird and mammal species. Once dominant tree and shrub species have fallen to introduced pests or are failing to regenerate. Just as scientists are beginning to appreciate more of the unique features of old-growth forests (Chapter 2), they are slipping away and becoming ecologically dysfunctional. The general public perceives very few consequences of these losses. An increasingly urban population and continuing declines in the scattered remnants of old growth suggest that fewer people every year will have the opportunity to experience old growth. Thus, we face the specter that today's and future generations will have little basis or concern for defending or restoring old-growth forests.

Many would argue that the early settlers who cleared most of the land could not afford the luxury of conserving their forests, preoccupied as they were with survival. European colonizers of a huge continent cloaked in primary forest could not imagine exhausting these forests or depleting the plants and animals they contained. For 200 years, the felling of forests and associated losses of wildlife occurred slowly relative to human life spans. Only at the end of the nineteenth century, after accelerated cutting had removed most eastern and midwestern forests, after game birds and mammals had become scarce, and after proliferating technology and transportation systems drove several species to the brink of extinction and threatened our final primeval forests, did public outcry arise to limit logging and protect wildlife. Still later came concerns to protect watersheds, re-vegetate stripped lands, and expand state and national parks, forests, and wildlife refuges.

This simplified view of nineteenth-century development obscures many of the social and commercial forces that favored agricultural settlement and deforestation (Cronon 1983, 1991). It also misleads many into believing that we have learned enough from our past mistakes to halt and reverse the destructive consequences of unbridled resource use and species declines. Early conservation measures did accomplish much in protecting soils, vestiges of remaining forests, and particular species directly threatened by overhunting. Their success, however, obscures the nature and extent of current threats to diversity and the profound failures we face as increasing numbers of species slide toward oblivion.

In this chapter we briefly review the sweep of biological change settlement has brought to our forested landscapes. We consider this historical background essential for appreciating the number and severity of these changes and for assessing how patterns of human disturbance threatened, and continue to threaten, diversity via the mechanisms described in Part II. We begin by examining how little remains of our original primary forests and where these remnants are located. We next examine how logging and development spread across the continent, altering landscapes and restructuring biological communities. Finally, we consider how the original primary forests differ from the secondary forests that succeeded them and the prospects this presents for restoring old-growth forests.

Readers interested in further information on the distribution and biotic history of forests in eastern North America should consult Braun (1950). Cronon (1983, 1991) provides insightful and informative ecological and economic histories of New England and the upper Midwest. Williams (1989) provides a detailed analysis of how logging developed geographically across the continent in accord with changes in technology. For readers with further interests in how conservation grew under the leadership of John Muir and his successors, we recommend Fox (1981). Those interested in how forestry and the U.S. Forest Service developed should consult Pinchot (1947), Frome (1962), Shepherd (1975), Steen (1976, 1984), and Clary (1986).

How Much Primary Forest Is Left?

Surprisingly few estimates exist of how much primary, or virgin, forest remains in the U.S. (Crumpacker et al. 1988). While no map has yet been drawn to show where these forests remain in the late twentieth century, several authors have recently ventured estimates of their extent. Postel and Ryan (1991) conservatively estimate that 85% of the primary forests had disappeared by the late 1980s. Other estimates range higher, to 90% (World Resources Institute 1992) or even 95–98% (Findley 1990).

Most of the eastern forests had been cut at least once by 1920 (Fig. 1-1).

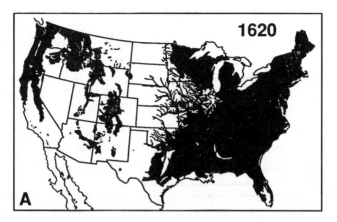

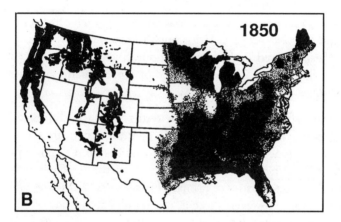

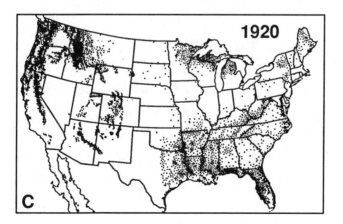

Fig. 1-1. Maps showing the amount of primary (virgin) forests remaining in the continental U.S. by 1620 (A), 1850 (B), and 1920 (C). [Figure from Greeley (1925).]

Today, less than 1% (Allen and Jackson 1992) and perhaps as little as 0.05%
(Leverett 1993) remains of the once expansive eastern virgin forests. Yet even
if we retain only an average of one part in two thousand of our original
eastern forests, the total acreage sums to nearly a million acres (400,000
hectares). Most of these stands are small and inconspicuous (Fig. 1-2),
leading some to deny their importance in forest conservation efforts generally. We believe instead that these remnants represent a legacy of significant
conservation value due to their scarcity on the landscape and potential for
helping to restore old-growth patterns and processes at larger scales
(Chapter 11).

This book focuses on the dichotomy that exists between primary and secondary forests. In keeping with Davis (1993), we follow the definition of primary forest provided by Duffy and Meier (1992):

> Forests that have never been clear cut and that have little or no evi-
> dence of past human activity. Such forests may have been grazed,
> . . . experienced limited exploitation of valuable tree species, and
> their floors may have been burned by Amerindians and European
> colonists.

Under this definition we will use the terms "primary," "original," and "old
growth" synonymously in this book. In contrast, secondary forests are those
that have developed after the previous forest has been extensively logged or
clearcut. Given the passage of long periods of time, secondary forests may
take on the characteristics of primary forests so as to become indistinguish-
able from them. However, a recent, much publicized study of old secondary
forests in the Appalachians suggests that 80 years appear insufficient for this
process to allow convergence (Duffy and Meier 1992).

Biotic Changes before Columbus

Ecological change has swept and shuffled the plant communities of North
America repeatedly over the past two million years. Climatic oscillations
brought glaciers from the North and dry winds from the Southwest. These
shifts apparently occurred slowly enough for most plant and animal species to
persist by shifting their distribution continuously (Davis 1965, 1969;
Delcourt et al. 1983; Peters and Darling 1985). In addition, refugia always
existed to provide appropriate climatic conditions and habitats at least somewhere in the landscape.

The arrival of humans radically altered the rate and nature of ecological
change. Peoples from Asia crossed into the Americas via the Bering Sea land
bridge by 11,000 years ago, bringing with them hunting techniques and

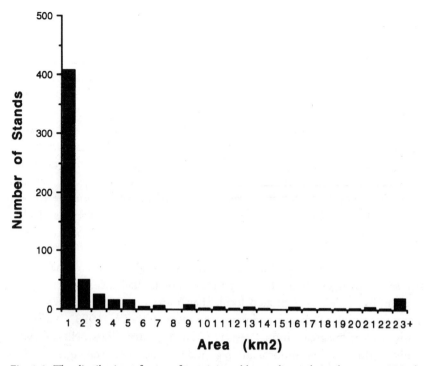

Fig. 1-2. The distribution of areas of remaining old-growth stands in the eastern U.S. described in Davis (1993). Note that 408 of the 528 stands listed are smaller than 100 hectares.

weapons technology that the continent's biota had never experienced. Almost immediately (in geological terms), most of the continent's largest herbivores (camelids, horses, mastodons, giant ground sloths, giant beaver, gomphotheres, etc.), predators (e.g., dire wolves and saber-toothed tigers), and scavengers (e.g., giant condors) went extinct. Many paleontologists believe that people, directly via hunting or indirectly via their cascading effects on other animals, precipitated these cataclysmic extinctions (Martin and Klein 1984). Having not coevolved with coordinated groups of neolithic hunters, these animals were ill-equipped to tolerate their depredations. This interpretation is bolstered by the observation that similar waves of extinction have repeatedly coincided with the arrival of humans onto other continents (Australia) and islands (Madagascar, Hawaii). Thus, humans have exerted potent ecological effects for many thousands of years on this continent, greatly modifying American landscapes via hunting, fire, and perhaps other activities. Their impacts precede and anticipate the even more potent recent impacts of industrial technology and agriculture.

European Colonization

Early European colonists found a continent full of game and majestic timber that dwarfed remaining forests in most of their home continent. While wildlife and abundant firewood may have been the privilege of the wealthy in Europe, any early American could feast on wild game and enjoy a roaring fire. Given such bounty and aspirations, it is hardly surprising that the colonists were preoccupied with clearing the land for farms and obtaining what useful materials they could from their forests. Indeed, wilderness forests often represented danger and an obstacle to be overcome by domesticating the landscape (Cronon 1983; Merchant 1989). While their use of game, timber, and land may seem profligate in retrospect, within the context of the time it simply reflected a function of plentiful supply and limited demand. Today, many visitors from China and parts of Europe view use of lumber and paper in the U.S. as similarly profligate.

In "settling" North America, the Europeans initiated an even more extensive set of changes on the landscape than Indians had. Most immediately, hunting by settlers precipitated local, then more extensive, extirpations of many remaining predatory and game species (Gates et al. 1983). They also began to clear the great virgin forests to plant crops, tend livestock, and eventually produce timber for local and regional markets. These changes, accelerated by population growth, industry, and silviculture, drastically altered the appearance and character of the land as native forest and prairie communities dwindled.

We next consider effects on wildlife and forests sequentially, emphasizing the upper Midwest to make our descriptions more detailed and concrete. By discussing these topics separately, we do not mean to imply that logging had effects clearly separable from overhunting. The trap, the rifle, and connections to distant settlements and markets had direct and profound effects on populations of our larger mammals. Nevertheless, it would be a mistake to ignore the roles timber cutting played in fostering the spread of settlements and the application of these tools (Cronon 1983, 1991). Logging drew settlers into new regions, provided markets for the game and furs, and contributed to the need for a transportation network and dams. In turn, roads and railroads provided access to the forests for hunters and trappers while increasing edge and open habitats.

Changes in Wildlife

Hunting and trapping followed the influx of European settlers, quickly depleting populations of predators, fur bearers, and the larger game species. Woodland bison and elk were extirpated by the latter nineteenth century, echoing the earlier losses attributed to Pleistocene hunters. Hunting and habitat changes eliminated woodland caribou and moose by about 1900 in Wisconsin and Michigan, while a few caribou survived until 1939 in Minnesota (Gates et al. 1983). These species depended on lowland northern forests and probably responded as much to the widespread fires as to logging. Caribou, moose, and elk still persist in Ontario, and moose continue to thrive on Isle Royale after colonizing this island in Lake Superior earlier this century.

Although white-tailed deer populations also were decimated in the upper Midwest by 1890, their high potential rate of increase coupled with subsequent protective hunting regulations allowed them to make a spectacular comeback. Responding to the abundant forage provided by the pervasive early successional habitats of the era, deer populations quickly expanded, surging farther north than ever before. Densities soared to record levels by the 1930s and 1940s, prompting warnings by, among others, Aldo Leopold (1943), that the herd should be reduced or forest tree seedling regeneration would fail. Leopold, the father of wildlife ecology, continued to lobby for strict regulation of the deer herd until his death.

Larger ungulates suffer especially under hunting pressure because they have low intrinsic rates of population increase. (Similar levels of hunting pressure have far less effect on deer because they mature faster than caribou or moose.) Top carnivores typically share this trait of low reproductive rates. Their reduced density and large home-range needs make them especially vulnerable to overhunting. Cougar (mountain lions) were hunted out of Michigan by about 1850, Wisconsin by 1884, and Minnesota by about 1900 (Gates et al. 1983). Canadian lynx also disappeared. Wolverines persist only as mascots in Michigan, having disappeared from that state and Wisconsin by the end of the nineteenth century and from Minnesota by 1918.

Although timber wolves originally ranged across almost all of North America, they have now been reduced to two enclaves in the lower 48: northern Minnesota and the northern Rockies. Wolves are beginning to spread back into the central Rockies and northern Wisconsin and Michigan, but only as scattered individuals or packs. A population colonized Isle Royale during the winter of 1946–47, grew quickly feeding on the many moose, but has recently declined to 12 individuals, reflecting an infection by parvovirus and possibly the genetic consequence of close inbreeding (Allen 1979; Mech 1966; Wayne et al. 1991). This population's looming extinction serves to warn us of the hazards other small populations face.

Fishers (large forest-dwelling weasels), trapped out of Wisconsin by 1927, were reintroduced (Powell 1982; Pils 1983) and now number approximately 6000 in Wisconsin (B. A. Kohn, pers. commun.). This species depends on

large snags and tree cavities, making it likely that logging contributed to its
demise or slows its recovery. Pine martens were also extirpated in Michigan
and Wisconsin then reintroduced, but they remain scarce and have not re-
covered to the same degree as fishers. Fur bearing mammals with higher rates
of increase, such as muskrat and mink, persisted even in the face of intense
trapping. Bobcats were considered vermin worth a bounty until 1963 in
Wisconsin and became subject to sport hunting in 1973. The Department of
Natural Resources (DNR) was petitioned in 1990 to protect the bobcat as a
threatened species, and the total Wisconsin population is now estimated at
1500 (W. A. Creed, pers. commun.). Ironically, abundant fishers may be
preying on bobcat kittens, contributing to their scarcity (J. Gilbert, pers.
commun.).

Despite being both abundant and prolific, beaver were almost extirpated
by about 1900 before they became protected (Schorger 1965). The young
forest that succeeded logging and fires was empty of wolves but full of favored
beaver forage (e.g., aspen and paper birch), allowing populations to bounce
back and even exceed presettlement densities (Gates et al. 1983). They now
pose dilemmas for forest managers in that they degrade high-quality trout
streams and damage some timber, yet provide dispersed wetland habitat that
support many other species. In contrast to the declines of most large mammals, rodents like the thirteen-lined ground squirrel and gray squirrel soared
in abundance and extended their ranges northward (Gates et al. 1983).

Fish populations also responded sensitively to logging and settlement.
Streams warmed by the loss of forest cover and siltation fostered the invasions
of exotic fish, like the parasitic sea lamprey and alewife, made possible by the
construction of the Welland Canal around Niagara Falls in 1829 and the Soo
Locks in 1855. These introductions, combined with over fishing on the
Great Lakes, precipitated the collapse of what had been the world's greatest
freshwater fishery. Mill wastes and conversion of wetlands to fields or pasture
lands further degraded the quality of many streams and rivers. Atlantic
salmon, lake trout, and massive sturgeon were eliminated as commercial
species one by one from each of the Great Lakes. With the massive disruption
of prey–predator relationships, many species went extinct, including larger
species of deepwater chub.

The mixed conifer-hardwood forests of the upper Midwest support bird
communities among the most diverse in North America (Gates et al. 1983).
These species also responded strongly to logging, with deep forest species
declining while edge species increased greatly in abundance (see Terborgh
1989; Chapter 5). Once scarce opportunistic predators like raccoons and
blue jays and brown-headed cowbirds (a nest parasite) also soared in abundance, reducing nesting success for many neotropical migrant songbirds.

Waterfowl and other game birds declined greatly in abundance until effective hunting regulations were established. Again, it was the species with special nesting requirements and slow rates of population increase that suffered the most. Once thought to be the most abundant bird species in North America, the passenger pigeon's colonial nesting habit made it both vulnerable to market hunters communicating via telegraph and sensitive to disruption of their nesting colonies (Blockstein and Tordoff 1985). Killed in astronomic numbers, this species dwindled to insignificant numbers by the turn of the century.

Wildlife protection has historically been equated with game laws that restrict hunting seasons or harvest. Game species such as wild turkey have now been successfully reintroduced or are regularly stocked. In contrast, other, less-popular or more demanding species (e.g., wolves, grizzly bears, and prairie chickens) have been slow to repopulate their original range or continue to decline. Most nongame species today are threatened by the slow but insidious indirect effects of habitat loss and fragmentation that game protection laws do little to address. As our primary forests have been logged and replaced by radically altered agricultural or silvicultural systems, some species have been robbed of their primary habitats while other, weedier species have thrived. Because the threats posed by such changes in the landscape and ecological interactions are diffuse and often emerge slowly, they are far less apparent to the public and harder for scientists to assess. Nevertheless, they are ultimately just as deadly as arrows or bullets and deserve our careful attention.

From First Cut to Clear Cut

Forest cutting increased rapidly through the latter half of the nineteenth and early twentieth centuries in response to developing technology, transportation systems, and access to markets. In this section, we review how logging followed (and sometimes led) patterns of European settlement across the U.S., sweeping down out of New England, through the Midwest, and on to the South and far West. We then review how our forests in the upper Midwest have changed in response to this logging and subsequent patterns of forestry and the lessons these changes hold for the Northwest.

The Northeast

New England loggers initially focused their attention on the tall, straight-grained white pines. Although smaller than the giant sequoias and redwoods of the Pacific slope, mature white pines grew larger and longer than any of the other trees, reaching diameters up to 6 feet and heights over 200 feet (Sargent

1933). These trees produced near-ideal ship's masts, particularly in relation to those produced from the depleted European forests. Starting with the charter from William and Mary creating the Province of Massachusetts Bay in 1691, the best trees were reserved for the Royal Navy. The "broad arrow policy" (named after the blazes on reserved trees) restricted the cutting of all pines greater than 24 inches in diameter and was extended to include all of New England, New York, and New Jersey by 1710 (Allen and Sharpe 1960). Contention over these valuable trees represented a significant element in the growing resentment of the colonies toward England that sparked the American Revolution.

Following extraction of the best white pines, primary forests in the Northeast were often felled to provide firewood or clear the land for agriculture. As many of the nearby original forests were logged, wood resources became scarcer and had to be brought from increasing distances, extending logging activities deeper into the mountains of New England and the Appalachians. This scarcity also prompted early attempts at conservation (Cronon 1983). Overcutting in the seventeenth century led William Penn to order in 1681 that one acre should be left in forest for every five acres cleared (a fraction far exceeding that which is deemed appropriate today by many in the Pacific Northwest). While parks and commons were established in towns and cities, these areas were, by design, imitations of the controlled, human-dominated landscapes of Europe (although rarely so contrived as European parks).

Early sentiments for conservation also were expressed by a few intellectuals such as H. D. Thoreau on aesthetic and moral grounds. Such viewpoints were unusual, however, both because of the prevailing climate of expansionism and also, perhaps, because few had perspective on the scope of change occurring. Thoreau gained such perspective by reading William Wood's account of New England in 1633 and traveling to the remote and still wild Maine woods (Cronon 1983), just as Leopold later gained perspective on the degraded nature of southwestern landscapes by traveling into the Rió Gavilan area of Mexico (Meine 1988).

Despite these sentiments, only a few medium- to large-sized areas of primary forest remain in the Northeast, notably in the uplands of the Adirondacks and Catskill State Parks in upstate New York and the Nancy Brook area of New Hampshire in the White Mountains (Leverett 1993). Other remaining areas of primary forest occur as scattered small patches (Fig. 1-2), a testimony to the thoroughness of early loggers and the effects of extended dense settlement.

Interestingly, forests have returned to cover much of the Northeast, and many of these have now matured to the point where they represent a significant and controversial natural asset (Mitchell 1992). Red cedars, pines, oaks, and eventually beech and maple invaded the abandoned fields, restoring forest cover to the hills and mountains of New England. Although these second-growth forests differ considerably from primary forests (see below), they have succeeded in transforming the landscapes of states like Massachusetts, which went from only 20% forest cover in the early nineteenth century to about 80% forest cover today. Because many of these forests were felled when demand for wood grew around the time of World War I (Fig. 1-3A), they often represent third or fourth growth. Nevertheless, they have sparked contention among those favoring recreational and second-home development, the wood products industry (with vast holdings particularly in Maine), and environmental groups intent on conserving the forests' growing natural value. Despite changes in structure and losses in biodiversity, these forests now provide an invaluable opportunity to restore ecological standards that might approximate those of the original forests (Niering 1992; Chapter 10).

The Midwest

As superior agricultural lands became available through the nineteenth century, farmers exchanged their rocky pastures and fields in New England for the rich flat soils of the Midwest. Wherever growing seasons were long and soils were rich and level enough, native vegetation was completely converted to farmland. Peak timber production shifted first to Pennsylvania, Ohio, and Indiana, then to the North Woods of Michigan, Wisconsin, and Minnesota (Fig. 1-3A).

Original forests in the upper Midwest were composed primarily of mixed hardwoods, swamp conifers, and pine stands. The hemlock–hardwood association covered over 60% of northern Wisconsin (Curtis 1959) and was dominated by hemlock, sugar maple, and yellow birch with scattered clumps of white pine and other hardwoods like red oak. Within these primary forests, tip-up mounds, standing snags, and rotting logs were common, affording denning sites for wildlife and germination sites for many plant species. By holding moisture, rotting "nurse" logs acted as critical regeneration sites for seedlings of many trees, including white pine, yellow birch, and hemlock.

In the mid-nineteenth century, these vast virgin forests on cheap federal land adjacent to rivers and the Great Lakes fueled a logging boom in the upper Midwest. The tall white pines that characterized the forest for many of the early voyageurs and woodsmen were the first to be cut, both because the wood made excellent lumber and because it floated, permitting convenient transport via the rivers of the region (Ahlgren and Ahlgren 1984; Karamanski 1989). Giant white pines became so scarce in the region that they now earn

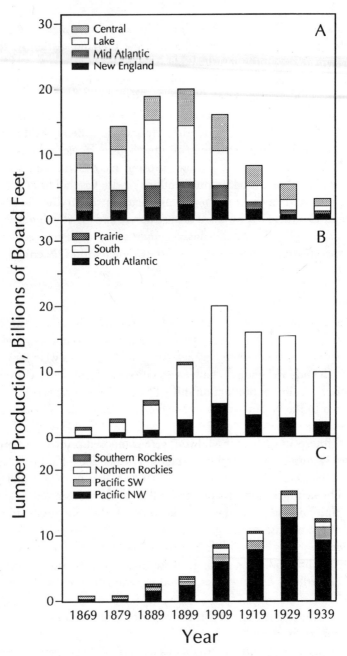

Fig. 1-3. Lumber production in the U.S., by region and decade, from 1869 through 1939: (A) Eastern and central regions, (B) southern regions, and (C) western regions of the U.S. Note the regional succession of boom and bust cycles. [Data from "Agricultural Conservation and Forestry Statistics" section of USDA Agricultural Yearbooks.]

plaques and individual titles. Trains, developing cities, and improved technology expanded lumber markets and extended timber harvesting throughout the region (Cronon 1991). As the pines were depleted, attention turned to eastern hemlock, whose bark supplied tannins for tanning leather. The growing pulp industry and demand for other forest products then led to the clearcutting of the remaining forests by the 1920s. The profusion of train lines in the region meant that rivers were no longer necessary to ship logs to mills and lumber to markets. Instead, mills could be located closer to the timber and lumber shipped directly to customers.

Between 1841 and 1891, the United States disposed of about two-thirds (over 600 million hectares) of the public domain, mostly to private individuals, corporations, and the states. Logging that had been practiced as a sustained activity in at least some areas of New England gave way to intensified and often abusive "cut and run" practices, where slash-covered logged lands were simply abandoned to avoid taxes. These episodes reflect the dominance of short-term economic forces in the absence of effective regulation.

The piles of slash (limbs, branches, and damaged trees) left behind after logging spawned an epidemic of forest fires through the late nineteenth and early twentieth centuries that had a profound and lasting effect on many of the forests that developed on these lands. These fires tended to be far more intense and correspondingly more destructive than the intermittent fires that had occurred previously. The Peshtigo fire in northeastern Wisconsin in 1871 devastated 50 square miles and killed 1500 people on the same day as the smaller but better known Chicago fire (Pyne 1982). Such extensive fires may have eradicated fire-sensitive, late-successional species with limited dispersal ability. Intense fires can sear the soil, oxidizing most of its organic matter and killing biologically important seeds, soil bacteria, and fungi. When such intense fires swept the Kingston Plains pineland in the Upper Peninsula of Michigan in the nineteenth century, the sandy soil was so thoroughly oxidized that no trees have grown there since.

Logging drastically altered the forested landscapes of the upper Midwest. Most conspicuously, early successional aspen, birches, and maples came in, relegating the original old-growth mixed hardwood forest to essentially token occurrence. Early successional forests and plantations of red and jack pines now dominate northern midwestern landscapes. Aspen, in particular, benefited from post-logging fires and intensive timber harvesting, increasing from roughly 1% before settlement to more than 30% of the public forests in the region today (USDA Forest Service 1986a,b). This species dominates industrial forest land and northern Minnesota to an even greater degree. Aspen is a preeminent early successional tree: its seeds are wind dispersed, its seedlings and saplings require high light and grow quickly, and it propagates clonally to occupy large continuous areas. Disturbances perpetuate and favor this species: following fire or cutting, it quickly root-sprouts. Aspen stands are managed on short (30–40-year) rotations to maintain their abundance and supply low-grade pulp for paper towels, toilet paper, disposable diapers, etc. Power companies in the region have also begun to use aspen as a biofuel to generate electricity. Such developments suggest that much of the landscape will remain in active management for aspen for the foreseeable future.

Changes in the fire regime also greatly altered plant communities. The intense slash fires eliminated seed sources for less vagile, fire-intolerant species. Subsequent fire suppression prevented intermittent ground fires that previously had consumed litter, reduced densities of shade-tolerant competitors, and created favorable conditions for the germination of species like northern red oak and white pine. Impacts were even greater in the sandy areas that had supported barrens vegetation of jack pine and associated grasses and forbs. All of these species declined after fires became infrequent.

The South

As loggers depleted the primary forests in the northern Midwest, they next turned their attention to the pine and deciduous river bottom forests of the South (Fig. 1-3B). Many of the pine lands had already been cut as agriculture had spread across the region; but cutting rapidly accelerated in the 1880s with an infusion of capital and technology from the North (Williams 1989). Thirty-six-million hectares of longleaf, shortleaf, loblolly, and slash pine were cut over by 1920, but only one-third of this was restocked with saw timber, leaving extensive scrub woods and barrens. With the depletion, mill towns vanished almost as quickly as they had sprouted, mirroring the booms and busts of the upper Midwest. Nevertheless, forestry persisted and even increased in importance. Repeated crops of cotton had depleted soil fertility in many areas, and the boll weevil further discouraged farming, allowing the timber industry to buy up large acreages during the Depression to establish pine plantations. Softwoods grow quickly in the South, and it was discovered by 1940 that pulp from young pines made excellent newsprint. Today, timber remains a big industry in the South, and pine plantations continue to dominate large acreages. As in the Northeast and Midwest, only small, scattered areas of primary forest were spared, notably in the Joyce Kilmer area of the Great Smoky Mountains National Park and the Congaree flood plain forest outside of Columbia, North Carolina (Fig. 1-2).

The West

After depletion of the North Woods, attention shifted west to forests in the Rocky Mountain region and the majestic conifers of the Pacific Northwest

(Fig. 1-3C). Here, most lands were unsuitable for agriculture, and difficulties of access and transportation limited extraction of timber resources to lowland areas near good harbors (Williams 1989). In addition, public sentiment had grown by the turn of the century to protect some areas as national parks and to retain other lands as national forests rather than disposing of them as had occurred elsewhere. Timber companies, however, appreciated the high value of the giant trees in this region, especially relative to their depleted forest lands in the Midwest. As policies to retain federal forest lands became clear, large timber companies like Weyerhaeuser moved quickly to buy remaining private tracts from railroad companies that had been granted huge areas in return for constructing their lines. Unlike in the Midwest, once these lands were acquired, they were not sold but rather were retained as industrial forests. These private industrial forests remain the base for much of the timber industry in the Northwest.

As elsewhere, timber was sold from the public forests in the Northwest at an increasing pace following World War II (Chapter 8). While many areas were cut, more primary forests still exist in the Pacific Northwest than elsewhere in the U.S., both because they have had the advantage of being remote and because the public has had time to insist on their protection. The survival of these majestic forests has also allowed for more detailed study. At the same time, their economic and natural value continue to grow, bringing these forests into a storm of conflict over how they should be managed (Ervin 1989).

Forests in the Pacific Northwest are complex and harbor many biologically unique species (Maser 1989; Norse 1989). Beyond containing the largest and some of the oldest trees, these forests are also home to an exceptional diversity of plant and animal life. Because of their maritime climate, the western slopes were never overrun by glaciers, allowing them to serve as refugia for plant and animal species in the region. The result is a high degree of endemism—a set of species restricted to the region that includes the Port Orford cedar, several salmonid fishes, the northern spotted owl, and many plant and insect endemics. Because many of these species thrive best in undisturbed, old-growth forests, management in their favor conflicts with efforts to harvest more of the remaining primary forests.

Of the original primary forests in the Northwest, only about 8% remain and only about half of these have been formally protected. In addition, the lands protected as Wilderness and National Parks occur mostly at high altitudes, in contrast to their plant and animal diversity, which declines conspicuously with altitude (Harris 1984). The result is that, although considerable public land remains protected in the West and Northwest, these lands do not include those that are most valuable from the point of view of conserving diversity. Economic and social pressures to complete the harvest of all available old growth have therefore run head-on into the Endangered Species Act and the diversity provisions of the National Forest Management Act (Chapter 12).

Summary

We have seen how a growing and industrializing society cut its way east to west through the forests of an entire continent in just three centuries. Our current forested landscapes differ radically from the forests that preceded them for millennia. Although concerns to conserve forests were expressed early in U.S. history, these sentiments were ineffective in slowing or preventing the logging of most eastern, southern, and midwestern forests. The mostly tiny remnants of these forests that remain provide valuable clues to their nature and the relationships that maintained their diversity. Only as the final western forests were being cut did the federal government act to protect larger primary forests by establishing national parks and forests. Controversy has reignited with the realization that these reserved lands are poorly situated and insufficient to protect many species dependent on old-growth forests in the region.

Early logging practices and the absence of effective wildlife protection brought tumultuous biological changes to these forests. Many of these changes, such as the loss of the great pineries and the extirpation of predators in the East, were obvious or dramatic. Great changes in forest age and composition have also occurred, but these have dropped from public concern as primary forests disappear from the landscape as a basis for comparison. Other changes continue that are far harder to discern but often profound in their influence on forest ecosystems. In the next chapter, we explore how far we still are from identifying the wide range of plant and animals species that dwell in our forests, let alone understanding their needs, dependencies, and interactions. Faced with such ignorance, we should ask ourselves when, and whether, we will ever understand the full consequences of our forest management on biological diversity.

2

Shadows in the Forest

Did significant losses of biodiversity in eastern forests occur only in response to drastic changes during past decades and centuries? And, does Chapter 1 suggest that eastern forest ecosystems are already so irreversibly altered from presettlement conditions that there's no reason to concern ourselves with further losses or forest restoration of any kind? Those who answer yes to these questions may feel there is little need to change current methods of forest and wildlife management. After all, if any serious damage to our biodiversity occurred only in the past, and we now have sufficient knowledge and tools to manage for the well-known and resilient species and interactions remaining, where's the problem?

In this chapter, we argue that despite previous losses and large strides in understanding biodiversity in temperate forest ecosystems, we have yet to comprehend enough to assume that present forest management strategies pose no threat to biodiversity in these ecosystems. This chapter, then, is about the magnitude of what we don't know about the biology of our forests, and the implications of this lack of knowledge.

The Search For Extra Terrestrial Life

In a recent essay, biologist Jared Diamond pointed out that we spent roughly one billion dollars on the Viking unmanned space missions to Mars, whose primary motivation was the discovery and description of extraterrestrial life (Diamond 1990). He mused over the fact that cumulative expenditures by our National Science Foundation for the discovery and classification of life on earth were "dwarfed" in comparison to this sum. The recent and extravagantly expensive failure of NASA's Mars Observer mission underscores Diamond's observation. Earthly explorations, and the subsequent description and classification of newly discovered biological organisms, were long considered by scientists and the public alike to be exciting and important endeavors, but there has been a decided change in attitudes during recent decades. Many scientists and the lay public now typically hold these descriptive studies of nature in less regard than they once did, albeit for different reasons. In the professional community, the taxonomists and systematists who carry out these studies are often viewed as hobbyist natural historians not really involved in the pursuit of important scientific knowledge. This disdain is expressed through reduced funding and professional opportunities for those pursuing the discovery and description of new organisms, resulting in a slow attrition of the ranks of professionals intimately familiar with the details of biodiversity of many groups of organisms (Disney 1989; Ehrenfeld 1989, 1993; Eldredge 1992).

As of late, taxonomists and systematists have managed to validate themselves in the eyes of some scientists by gathering data that can be used to explicitly "test" hypotheses about evolution and by employing new and powerful molecular tools (La Duke 1987; Luoma 1991a, 1992). This testing and rejecting of hypotheses is often considered to be the hallmark of science, separating the scorned "descriptive" methods of natural history from the favored "experimental" methods of science.

There is a more disturbing side to the issue, however, that goes beyond these valid arguments over scientific methodology. Many members of the public, and a disappointing number of scientists, do not support an expanded search for life on earth simply because they don't believe that anything interesting or significant remains to be discovered. From this point of view, the things still unknown to us are largely unimportant details to be figured out by technicians and are not a high priority when compared to other scientific pursuits. For many, the exploration of life on earth has lost its wonder.

The Tip of the Iceberg

It is true that few, if any, of the species yet to be discovered on earth will rival, in size and quantity of fur or feathers, the big finds of previous centuries. Yetis notwithstanding, most of what is big, terrestrial, and conspicuous (often through being common) has already been discovered. In 1992, a goatlike mammal, the pseudoryx, was discovered in a remote Vietnamese forest* (Basler 1993; Van Dung et al. 1993) 55 years after the last find of a big terrestrial mammal, an Indochinese ox called the kouprey (Diamond 1985a). New mammals and birds still turn up, but at a relatively slow rate. From 1940 to 1983, biologists discovered an average of less than one new genus of mammals per year, none of which was from the U.S. or Europe; and the same 43-year period shows an average discovery rate of five bird species every two years (Simpson 1984; Diamond 1985a). New species of mammals are now found at a rate of about five per year (R. Timm, pers. comm.).

If we turn our attention to groups that are less like our land-based, common, and relatively large selves, it quickly becomes apparent that our knowledge of life on earth is quite limited. For example, a new era of oceanic

*While this book was in press, yet another large mammal, the giant muntjac deer, was discovered in this same Vu Quang forest (*The New York Times,* April 26, 1994, p. B7).

exploration has just begun, bringing with it remarkable finds of new whales and giant-mouthed sharks (Diamond 1985b; Wilson 1992), deep-sea communities of sulfur- and methane-eating organisms heaped upon superhot volcanic vents and whale carcasses (Smith 1990; Broad 1993a), the world's deepest-dwelling plants (Littler et al. 1985) and microbes (Broad 1993b) and species representing previously unknown, extremely distinct groups of organisms such as the Loricifera (tiny "girdle-bearing" animals which dwell between grains of sand on the deep-sea floor—Raven and Johnson 1986), the Monoplacophora (single-shelled, deep-water relatives of clams—Margulis and Schwartz 1987), and the Concentricycloidea (sea-daisies, relatives of starfish—Nichols 1986). No one seriously doubts that far more discoveries await us in the oceanic depths.

Small and otherwise inconspicuous organisms are often the most poorly known, since we tend to overlook them unless we search carefully and specifically. For example, Loriciferan animals turned out to be fairly common and widespread once we realized they existed and learned how to search for them. In fact, similarly painstaking, focused studies of small organisms in terrestrial, not oceanic, habitats recently catalyzed a major change in thinking about the overall number of species on earth.

In the late 1970s Smithsonian researcher Terry Erwin and associates devised a method to make relatively complete collection of insects and other arthropods (spiders, mites, centipedes, etc.) from individual trees growing in forests of tropical Central and South America. During windless evening hours, a portable fogger was used to loft a cloud of insecticidal spray directly up and through the leafy crown of each tree to be sampled. The arthropods, relaxed in death, tumbled into traps distributed beneath the trees. The collection method worked as planned but the results were highly unexpected: the number of insect and other arthropod species falling from individual trees was astronomically greater than anticipated, and each kind of tree sampled seemed to yield unique insect species. Through continued sampling, and by extrapolation of his findings to other species of tropical trees, Erwin estimated that there may well be 30 million extant species of arthropods (Irwin 1982, 1988). This number is highly contested, and recent, alternative estimates range from "below 5" to over 10 million arthropod species alone (Gaston 1991a,b; Irwin 1991; Hodkinson 1992; Yoon 1993). In contrast, however, only about 875,000 arthropod species worldwide are formally recognized at present (Wilson 1992). A comparison of these estimates clearly suggests that the majority, perhaps a vast majority, of the world's arthropod species remain undescribed. Within the community of conservation biologists and systematists, Erwin's studies catalyzed a shift of attention from the relative security of what is known about arthropods to a less comfortable awareness of the overwhelming magnitude of our ignorance.

A similarly extreme example comes from the even smaller realm of bacteria. Fewer than 4000 species are formally recognized by microbiologists (Trüper 1992; Wilson 1992). But, because many bacteria won't reveal their presence by growing on the agar plates used for standard assays, it has long been suspected that their numbers are far greater. Recent studies in Norway directly examined bacterial DNA in an attempt to avoid the pitfalls of standard assays and found between 4000 and 5000 species of bacteria in a single gram of beech-forest soil. A second gram examined from sediments in shallow waters off the Norwegian coast yielded as many species, but virtually all of them different from the forest sample. These and other recent investigations suggest that the number of bacterial species worldwide may be 100 to 1000 times that previously thought (Trüper 1992; Wilson 1992; Pedrós-Alió 1993). Impressive examples can also be cited for our ignorance of algae (of which it is estimated that less than 4% are known and described—Andersen 1992), fungi (Oberwinkler 1992), protozoans (Bull 1992; Vickerman 1992), and viruses (Bull 1992).

Overall, we can only guess at the number of species that exist on earth. Somewhere around 1.4 million are known, having at least been formally described and given a name; but the overall total probably falls somewhere between 10 and 100 million (May 1986, 1988; Wilson 1992). Put another way, we can only name somewhere between 1 and 15% of the organisms on the planet and have good ecological information on a far smaller subset.

Local Lacunae

Accounts of bizarre deep-sea denizens, bugs in tropical trees, and bacteria in Norwegian soil may appear interesting but remote to many readers. Perhaps biodiversity, peculiar and engrossing as it may be, is something to be experienced mostly through nature programs on the TV?

In 1991, the Illinois Natural History Survey, one of the premier state natural history agencies, provided estimates of the number of species thought to exist in Illinois. Three groups were considered intractable—bacteria, nematodes (roundworms), and protozoans (single-celled animals)—so these very speciose groups are not reflected in the overall, admittedly crude estimate: 53,754 species, minimum (Post 1991). A recent, very conservative estimate for Missouri suggested a minimum of 25,000 species for that state (Nigh et al. 1992).

The biotas of Illinois and Wisconsin are relatively well known, at least in comparison to places like Borneo and Colombia. Not only can we name all of the birds, mammals, reptiles, and amphibians that occur, but reasonably good ecological data compiled through years of study by wildlife biologists, academicians, and amateur naturalists exist for each.

Plant lists continue to grow for two reasons. In Wisconsin, we add one or two species of vascular plants each year, on average, either because they have been here all along and we finally managed to detect them or, more likely, because they are native to somewhere else on the globe and have recently been transported into our lakes, fields, roadsides, and clearcuts. The latter species, called exotics, have constituted the bulk of the new records for at least the last few decades and will continue to inflate the state plant list in the foreseeable future. New discoveries of plant species native to Wisconsin (i.e., nonexotics) will also occur through time, but at a much lower rate.

Other plant groups are not as well known, as suggested by recent field studies. A single afternoon's collecting in several old-growth stands by Wisconsin lichenologist John Thomson yielded species previously not known to inhabit these areas but characteristic of old-growth forest communities in the Pacific Northwest and Europe (Thomson 1990). A Minnesota lichenologist, collecting over a three-week period in the Rainbow Lake Wilderness Area of the Chequamegon National Forest in northwestern Wisconsin, found 21 species of lichens new to the state, comprising 11% of the species he collected in this 2000-hectare reserve (Whetmore 1993). Both of these studies suggest that additional surveys, including thorough and systematic studies of uncommon microhabitats, such as the upper branches of trees in old-growth stands, will significantly increase the number of lichens known from these forests.

For other groups of organisms, especially at smaller biotic scales, we know far less, even in forest stands superficially quite familiar to us. In Wisconsin, we know almost nothing about the insects associated with the lichens growing on the bark of old trees or about the identity of the insects, mites, and other invertebrates in forest soils. And, unable to name them, we cannot store or retrieve information about their ecologies and interactions useful for management purposes except in the most general terms.

Pioneering studies of soil invertebrates in old-growth forest stands of the Pacific Northwest by Andrew Moldenke and colleagues yielded upwards of 200 to 250 species per square meter, many of them undescribed (Moldenke 1990; Luoma 1991b). That is, they were not merely new to that specific forest, or to the state of Oregon, but new species never before known to exist on earth. Like the studies of tree-dwelling tropical insects, and of the Loriciferans tenaciously clinging to grains of sand thousands of meters beneath the sea's surface, Moldenke's studies required the development of a new method of sampling. Combined with his intent to carefully scrutinize the diversity of invertebrate species inhabiting these forest soils, this new method yielded far more species than anyone expected. When added to the number of arthropods living above ground, these investigators estimated a total number of approximately 8000 species within their 6400-hectare study site (Lattin 1990; Beard 1991). To our knowledge, no comparable studies have been conducted in eastern forests, but it is reasonable to expect that they might yield results of the same order of magnitude, especially if old-growth forest stands were included.

The Relevance of Ignorance

There are many reasons to value and conserve biodiversity, including economic, intellectual, spiritual, and technological. Excellent discussions of these exist elsewhere in very articulate and complete forms (Leopold 1953; Iltis 1970; Ehrenfeld 1978, 1993; Ehrlich and Ehrlich 1981; Ehrlich and Wilson 1991; Wilson 1984, 1988, 1992; Manes 1990; Foreman 1991a; Ricklefs 1993). Rather than review these reasons here, we consider instead how our ignorance of the components of biodiversity is relevant to forest management in North America.

Concepts of nature and wildlife shape management practices

Many resource professionals, especially those working with relatively well-known aspects of ecosystems, do not spend much time thinking about poorly known species and their interactions. Yet, a failure to appreciate what we do not know about natural systems can manifest itself in decisions that damage elements of the hidden biota. Even when biodiversity is professed as a management goal, concepts of the natural world consciously and unconsciously shape decisions in a way that will favor certain elements of the ecosystem and disfavor others. We manage only for what we think of as being "out there" in nature.

This is perhaps most clearly manifested in the way that the term "wildlife" is used. The term commonly appears in the literature of forest management and in educational materials provided to the general public, but it is not used in a consistent manner. The technical literature usually equates wildlife with game, or with the broader set of species that includes all animals with backbones (Chapter 5). In contrast, booklets and pamphlets for public consumption often contain pictures not only of these groups of organisms but also of butterflies, plants, and mushrooms, implying that wildlife is defined to be a much more inclusive group.

Forest management proposals for local road and timber projects often proclaim the benefits to wildlife. But, one must ask, which wildlife? These projects are clearly good for many game animals, as well as some other verte-brates. But if wildlife is defined more broadly to include all organisms living in that place, are such positive and unqualified pronouncements warranted? The conviction that active management to achieve younger, more vigorously growing, "healthier" forest stands is good for all wildlife, however poorly sub-stantiated, has become deeply ingrained in the thinking of many (Chapters 5, 8, and 13). Proponents of this point of view have been encouraged to con-sider the effects of traditional management techniques on populations of a broader array of species than are usually considered (e.g., Temple 1986; Tilghman and Evans 1986), since these effects are often negative or of du-bious benefit (Perry et al. 1989; Franklin 1993).

We do not wish to imply that there is anything inherently wrong with managing for game species, or for managing for other desirable mammals, birds, or fish. We suggest only that managers keep in mind the huge propor-tion of the biota that we know little about and devote more attention to the broad spectrum of wildlife affected by current managment practices.

Ecosystem function depends on unseen organisms

Too great an emphasis on familiar elements of regional ecosystems may di-minish our ability to detect and respond to changes in populations of poorly known species that are central to nutrient cycling, long-term site produc-tivity, and responses of other ecosystem components to pathogens and exotics (Ehrlich 1988; Ehrlich and Wilson 1991).

Invertebrates and fungi of forest litter and soils come to mind because of their incredible abundance and diversity (e.g., orbatid mites in Oregon, with 250,000 individuals representing 75 to 100 species per square meter of forest floor—Moldenke and Lattin 1990a). Similarly, the slender underground threads of hundreds of species of mycorrhizal fungi weave complex networks among decaying litter, wood, and other organic materials, shunting some of these nutrients to the roots of trees. Individual trees often depend on scores of mycorrhizal species, each with different ecological characteristics and re-sponses to stress (Oberwinkler 1992; Read et al. 1992).

We have only begun to study the complex interactions of soil arthropods and fungi and their long-term consequences (Cromack et al. 1988; Jaenike 1991; Shaw et al. 1990; Klopatek et al. 1992; Lussenhop 1992; Kremen et al. 1993). Nor do we understand how traditional methods of forest manage-ment, such as wildlife openings, clearcuts, winter selective cutting, and the re-duction of acreage of old growth, will affect these hidden elements of forest ecosystems. It is, however, beyond doubt that all of these activities can and do significantly alter ecological relationships (Schowalter et al. 1981; Schowalter 1989; Moldenke and Lattin 1990b; Shaw et al. 1990; Niemelä et al. 1993).

Management for wildlife in the narrow sense affects wildlife in the broad sense, whether intended or not.

In the examples just discussed, many species interact in ways as yet poorly understood to provide benefits and stability to familiar elements of the ecosystem, producing healthy trees, fertile soil, good forage for deer, and so on. Yet, even the tiniest individual organisms sometimes flare out from this biological tapestry to alter temperate ecosystems in dramatic and long-lasting ways. Infestations of exotic organisms like chestnut blight, gypsy moths, hemlock woolly adelgids, and beech scale insects grab our attention because of the striking damage they do to our forests (Burkman et al. 1993; Campbell and Schlarbaum 1994). The fundamental cause of these infestations in many instances is the exposure of organisms (particular trees in this case) to novel pathogens (insects and associated fungi) to which they have not been previously exposed and, therefore, have not been able to adapt genetically. This line of reasoning is easy to grasp, since the pathogens appear to have only recently arrived from far-off places. What is much more difficult to explain, however, is why these particular organisms, and not others of the thousands of new arrivals, have become virulent. Or how the millions of native organisms, each of which might be a serious pest if transported to places like Australia or Hawaii, interact here at home in a way that counteracts the community-level effects of such extreme and abrupt changes in the abundance of individual species. Anyone who doubts that knowledge of tiny, individual components is relevant to ecosystem health, or to the welfare of individual species of particular value, should consider the tremendous uncertainty surrounding the mechanisms governing the origin and initial spread of AIDS (HIVs) and even more lethal viruses (Preston 1992; Levins et al. 1994).

The role of pathogenic microorganisms has long been considered in game animals and, more recently, in the context of conservation efforts at the community level (Dobson and May 1986). Future studies will undoubtedly approach the subject by investigating undesirable pathogenic effects on individual species of value to us, including favored nongame vertebrate species and commercially valuable plants (Stanosz and Patton 1987; Hansen and Littke 1993). Beyond this lies the question of the way current microbial assemblages, and the genetics of other forest organisms, interact to produce the regimes of disease, decay, and competition that determine how our forest ecosystems look and function.

How to "fix" damaged ecosystems in view of our ignorance

Conservation biology, by its own admission, is saddled with the dilemma of choosing between active versus passive management. At what point do we know enough to jump in and "do things" to heal or constructively alter ecosystems? Our own bias is to err on the side of passivity unless and until we have a reasonably good understanding of the results of active, interventive management (Chapter 11). Here we only wish to stress the degree to which our knowledge of the biota of temperate forests significantly limits our ability to wisely or responsibly restore and manage ecosystems.

One of Wisconsin's rare orchids, *Calypso bulbosa*, grows in cold, mossy swamps dominated by old white cedar trees in the northern third of the state. At present, we can only describe its preferred microhabitats in a crude way. We cannot specify how best to provide for long-term calypso habitat. Nearby white cedar stands that look like prime habitat do not harbor these orchids. Are unseen but critical biotic elements missing: mycorrhizal associates (critical to many orchids) or insects that somehow benefit these plants? Perhaps these orchids are not now present because of pathogens built up on past populations at these sites, lingering to prevent successful colonization by the occasional rain of tiny orchid seeds. Or perhaps the seeds simply never get there. We don't have a clue.

Calypso is not an isolated example. Restoration efforts of relatively common elements of our temperate forests, such as eastern hemlock, red oak, and white pine, are clouded by unresolved issues over the roles played by herbivores such as deer, soil conditions, and the presence or absence of mycorrhizal associates relative to the establishment and spread of these species. We know that we can plant these species in pots, surround them with wire fence or Tubex, and get them to grow successfully. What we don't know is why they are not doing well on their own under current ecological conditions.

Our attempts at ecosystem management, seen in this light, evoke the image of an ambulance team attending the bloodied crew of a crashed flying saucer. We desperately wish to help the victims but we are not quite certain what to do. Despite our advanced training and skills, what we see is bewilderingly alien to us. Is this orifice supposed to be drawing air or should we consider it a lesion to be plugged? This looks like a heart but should we administer CPR or will it start again on its own—or is it really a gizzard? How, then, should we administer to the remaining old-growth habitats and similar biotic debris that have thus far survived the crash of the North American biota against the wall of post-Columbian technology?

Summary

Wolverines no longer galumph across the Wisconsin landscape, nor do elk or caribou roam here and feel spruce needles comb their flanks. What other elements of biodiversity characteristic of our undisturbed presettlement forests survived the intense logging of the past described in Chapter 1? We know that the old-growth lichen *Lobaria pulmonaria* still resides on the bark of old yellow birch trees in scattered places in northern Wisconsin (Thomson 1990; Alverson, pers. obs.). But we cannot yet say whether populations of mites and beetles associated with this lichen in presettlement old-growth stands still occur here—no one has looked. It is encouraging, however, to know that when investigators carefully scrutinized the treetops, bark, leaf surfaces, rotten logs, and soil layers of forests in the western U.S. they found more than they could have imagined. There's every reason to think that the components of diversity in forests of Wisconsin, New York, and North Carolina also number at least in the tens or hundreds of thousands but remain largely hidden from our view.

The task of revealing each and every component of biological diversity is beyond our intellectual and economic resources. With an increased awareness of the diversity of wildlife inhabiting our forests, the domain of our ignorance has grown much more quickly than that of our knowledge. For better or worse, we must count on, and plan for, how best to manage these forests in view of what we don't know.

Despite the hopelessness of obtaining complete biological inventories, we must learn much more about poorly known ecosystem components, especially abundant, species-rich groups like invertebrates and fungi. From such studies will emerge not only a better understanding of these individual parts, but also a more complete picture of how ecosystems function. Such information will allow us to better elucidate the effects of traditional and recent management strategies on forest biodiversity (Chapters 8–10). Are we currently practicing sustainable ecosystem management? Without more and better inventories of forest ecosystems, it is too early to tell.

3

The Myth of Heart's Content

Driving through the small town of Warren, Pennsylvania, one turns off the main street at a conspicuous memorial obelisk to proceed southward to Heart's Content, the famous old-growth stand. The memorial itself commemorates Warren County veterans and bears on its base the words "the bivouac of the dead" as the final verse of a short poem. Soon after this turn, the greenery of the Allegheny National Forest envelopes the roadway, and even with several curiosity stops one can stand before the thick wooden sign at the entrance to the Heart's Content trail well within an hour.

Spared from the ax because the Wheeler family didn't want to "slick off" the timber around their homestead, Heart's Content is a National Forest Scenic Area formed of two parcels dedicated in 1929 and 1931. The National Park Service subsequently designated the stand as a Registered National Natural Landmark in 1977. Hikers are greeted by an informational flyer available at the trailhead asserting that "as a National Natural Landmark the significant ecological and historical value of this area will be protected and preserved forever." The impressive canopy trees of this stand—hemlock, white pine, maple, and beech—are 200–400 years old. Yet, despite its formal designation as a biological showcase and its evocative and euphonious name, this 49-hectare old-growth stand has severe biological problems: lost species, sharp reductions in population sizes of other species (with likely decrease in their genetic diversity), and significant alteration of natural ecological processes during our lifetimes.

Like Wisconsin's forests, the 2.4 million hectares of old-growth forest that once comprised much of the Allegheny Plateau of Pennsylvania and western New York were home to the eastern gray wolf, elk, moose, and wolverine. All are now extirpated, with uncertain consequences for remnant old-growth stands like Heart's Content. Soon after Heart's Content's formal designation, floristic and vegetative surveys were conducted in it and nearby old-growth stands (Lutz 1930; Morey 1936; Hough 1936). A 1929 vegetation survey in this white pine–hemlock–hardwood stand tallied 83 herbaceous species; accompanying photos show dense understory growth. Today, however, the understory is sparse and many species such as hobblebush, red-berried elder, trillium, and wild sarsaparilla have been lost from the stand ("extirpated") or exist as a few, scattered remnant individuals. Some of the trees, notably hemlock, yellow birch, sugar maple, and cucumber magnolia, have failed to regenerate, even though one would expect the normal course of ecological succession to slowly move the stand toward an old-growth mixture of these shade-tolerant species (Whitney 1984). Instead, a plethora of beech sprouts in the understory suggest that the stand may unnaturally become a nearly pure stand of that species.

These drastic changes in stand composition cannot be attributed to global warming, nor are they peculiar just to Heart's Content. Similar changes have taken place in other old-growth stands on the Allegheny Plateau (Smith 1989), and many can be attributed to huge population increases of white-tailed deer:

> Any attempt to maintain a limited tract of forest in a natural condition is subject to the hazards of a man-caused build-up of herbivorous animal populations in the large surrounding area. This upsetting of the natural balance could not occur if the entire Allegheny Plateau were in its original virgin condition. However, logging, extermination of predators, reintroduction of the white-tailed deer on game refuges, and protection of this species by game laws, have resulted in large deer populations and consequent hazards to tree growth. Unless some way is found to keep animal populations, such as deer, in balance in this particular forest stand, the objectives of scientific study, public education, and historic value will inevitably be lost. (Hough 1965, p. 373)

Deer populations of the Allegheny Plateau have grown from presettlement densities of about 4–8 per square kilometer to densities of 10–20 per square kilometer at present (deCalesta 1992). At the most recent population peak, in the mid-1970s, these forests were browsed by deer numbering 22–32 per square kilometer. Deer exclosures, small cages designed to keep deer from entering and browsing within experimental plots, erected within Heart's Content and other forest stands demonstrate the vast effect this increased browsing has on the floristic composition and the vegetative structure of these forests (Fig. 3-1) (Marquis 1981; deCalesta 1992; Horsley 1992, 1993).

These changes are not limited to the vegetation. Researchers are only now beginning to investigate the ways in which deer browsing, through alterations in understory structure and competition for food, affects populations of native mammals and birds (McShea and Rappole 1992; deCalesta 1994a,b). We cannot even begin to assess the effects of these changes in terms of invertebrates, fungi, and other types of organisms resident in Heart's Content, since no comprehensive inventories have been conducted. Yet, the "deer problem" is not easy to solve. As in Wisconsin, a powerful hunter lobby cam-

Fig. 3-1. Deer exclosure within the Heart's Content old-growth stand in the Allegheny National Forest in northwestern Pennsylvania. Protected from browsing by the deer-proof cage, yellow birch and other woody plants grow vigorously, in contrast to the sparse understory outside of the exclosure. [Photo: W. Alverson, 1992.]

paigns for abundant and widespread deer for sport and venison. Even more problematic is the fact that any biological solution to the deer problem lies not within these old-growth stands but in the surrounding landscape of young forests and fields that engender the high deer populations.

Heart's Content and nearby old-growth stands are vulnerable to two other obvious external threats. Exotic diseases such as chestnut blight and beech bark disease have removed chestnut, once a common tree in Heart's Content, and are now killing off most canopy beech from the stand (Whitney 1984; Alverson pers. obs. 1992). Heart's Content, because of its small size, is also highly vulnerable to catastrophic destruction. A single windstorm could easily devastate the entire stand in a few seconds, just as a brief and furious storm in 1985 flattened 400 hectares of the Tionesta old-growth stand to the east of Heart's Content (Fig. 3-2) (Mohlenbrock 1986; Peterson and Pickett 1991). The same storm also devastated small old-growth stands elsewhere in the Allegheny National Forest (Davis 1993).

Thus, a walk through Heart's Content oddly brings back the inscription on the memorial obelisk in Warren—"the bivouac of the dead." Many of the biological lineages present in Heart's Content have lived together or in close proximity for thousands of years. One of these lineages, eastern hemlock, now exists in old-growth stands like Heart's Content primarily as older trees

Fig. 3-2. Blowdown in the Tionesta Scenic Area of the Allegheny National Forest. A violent tornado struck the Tionesta on May 31, 1985, removing 400 of the 1600 hectares of old growth. [Photo: W. Alverson, 1987.]

whose seedlings and saplings are browsed to death by deer. These old hemlocks can be considered "living dead" from an evolutionary standpoint because, though still alive themselves, they have little hope of passing on their genetic heritage to the next generation. From the perspective of these hemlocks, the post-settlement timber harvests that have changed the face of the Allegheny Plateau must appear as a sort of blitzkrieg, a short and devastating war that left them and other survivors scattered and isolated, bivouacked in small, remnant old-growth stands surrounded by an often hostile matrix of young secondary forests and fields.

Future Forest Diversity

Assuming that we can provide for efficient and sustainable commodity production in our forests, how do we simultaneously protect biological diversity? Do we do so on the same or different areas? To what degree do we actively intercede with management?

Some may envision protected old-growth stands as nothing more than curious historic relics that imperfectly communicate the nature of forests in a bygone era. Showcase stands like Heart's Content are tolerated, since their overall acreage is quite small. But these stands are seen as largely irrelevant to a vision of the "working forest" that produces abundant forest products as we symbiotically husband its resources through judicious management. From this viewpoint, both forests and human caretakers benefit from the relationship, and forests can indefinitely prosper and remain "healthy" only through our care. Indeed, were the human caretakers withdrawn, not only would substantial assets of the forests be "wasted," but the forest itself would become inflicted with diseases and pests, and the trees and wildlife would be reduced in number and health.

A second point of view now becoming ascendant in contemporary forest management (Chapters 8–10), concedes that old-growth habitats such as Heart's Content possess regionally significant elements of biodiversity and proposes to maintain these elements through strategic, and often intensive, "active management." For example, one solution to the browsing problem at Heart's Content would be simply to erect a deer-proof fence around the entire old-growth stand. Similar measures could presumably be taken at other old-growth stands throughout the Allegheny National Forest, just as thousands of hectares of clearcuts on the Allegheny Plateau are now surrounded by deer-proof fences so that commercially valuable timber species can be successfully regenerated. Extending this vision into the future, forest managers would become increasingly sophisticated and knowledgeable and would attempt to address and correct all significant ecological threats to old-growth habitats and their constituent biota on a species by species basis. These forest managers would reintroduce locally extirpated species, closely monitor and control the dynamics of insect and fungal interactions, and actively mitigate subtle but insidious losses in genetic diversity. The ultimate goal, under this vision, is to actively create forest habitats that can produce a full range of commodity (lumber, fiber, etc.) and noncommodity (e.g., recreation, biodiversity) benefits on the same acreage.

A third point of view stands in contrast to the previous two in that it eschews active management in favor of a hands-off approach. Advocates of such a philosophy would argue that the ecological ills of Heart's Content will eventually correct themselves, or that these ills should be addressed through active management outside, not within, the old-growth stand. Advocates might envision the restoration of large forest tracts that would look much as forests did before European settlement. Detractors would claim that this is a scientifically and politically naive viewpoint, given the number and nature of external threats to Heart's Content, and would consider such restorative goals arbitrary and unimportant.

What, then, is the best approach to protect the diversity of Heart's

Content? In fact, no specific, long-term management plan exists for Heart's Content aside from general consideration as a special conservation management (MA-8) area in the Forest Plan, which covers a 200,000-hectare area (USDA Forest Service 1986c). Fencing of all or part of the stand has been discussed over the last few years but has been rejected for lack of resources (A. Irvine, pers. comm.). Management for Heart's Content will be addressed, in part, in a formal Plan Amendment to be proposed in 1994. This Plan Amendment will advocate a winding network of restored old-growth corridors across much of the Allegheny National Forest (B. Nelson, pers. comm.). Yet many questions persist. How will commodity production be coordinated with old-growth restoration in these corridors? Will the corridors, predominantly less than 2 kilometers wide, function as old growth themselves? To what degree will this network buffer existing old-growth stands against pernicious ecological factors in the surrounding landscape?

Summary

We need a clear, coherent vision of how best to maintain and restore biodiversity in temperate eastern forests. If there are unique components of biodiversity contained within remnant old-growth stands in the East, we must utilize the best scientific information available to predict how to provide for this diversity over time. This vision must also consider the need to produce other forest products, at least somewhere in the same forest landscape. In Part II we turn our attention to the scientific tools and principles with which we can formulate a biologically meaningful vision that recognizes but is not constrained by short-term economic consideration.

Ecological Mechanisms and Biotic Resources

LIKE FORESTRY, METEOROLOGY, WILDLIFE BIOLOGY, and epidemiology, conservation biology is a science of complex interactions that only roughly describes the present and imperfectly predicts the future. Although foresters can usually anticipate with some accuracy the composition and growth rates of young stands resulting from the harvest of old growth, we cannot yet predict how colonies of soil invertebrates or canopy lichens will respond to such cutting locally or across the landscape.

When confronted with uncertainty regarding an issue of major importance, scientists and planners often rely on models to predict the consequences of alternative courses of action. Scientific models are usually based on at least some empirical data but still vary in their conceptual validity or in the amount of supporting information they use. Importantly, models provide testable hypotheses with which we can explore the implications of our assumptions. We attempt to formulate them using the best scientific data available, then make predictions to see how well they are borne out in similar situations. If our predictions are wrong, we suspect our assumptions and either reformulate or junk the model.

We routinely accept the limitations of our models when trying to predict complex phenomena such as the spread of an epidemic or global weather patterns. For example, we can only imperfectly predict the origin and spread of new diseases, usually because we lack sufficient information regarding the microbe, its modes of transfer, and interactions with its host(s) (Brown and Lane 1992; Preston 1992; French et al. 1994). Similarly, meteorological models function well at some scales but still cannot accurately predict local or long-term weather phenomena (e.g., Will hail flatten these cornfields or those? What will the average humidity be within this forest canopy next year?).

We are faced with similar limitations regarding the models we use in managing forests. The science of wildlife biology has provided models describing how changes in vegetation affect deer populations. In general terms, we know that increasing the amount of young forest and open lands will, up to a point, increase the number of deer in a region (McCaffery 1986). However, because deer are relatively mobile animals that react in as yet poorly understood ways to changes in patterns of vegetation at the landscape level, these models are not accurate at smaller scales. In conserving biodiversity, one would like to predict how managing vegetation elsewhere in a region will affect the number of deer wintering in a particular old-growth conifer stand. Our current models, however, are not yet capable of predicting such details regarding deer–habitat relationships.

Foresters, wildlife biologists, and conservation biologists will all play important roles in conserving our native biota and the ecosystem processes they depend on within our forests. To that end, these groups of scientists must use or modify existing models from one another's disciplines, or create new models to better predict the welfare of all wildlife under various management scenarios. As our knowledge and predictive tools improve, we gain a clearer picture of how ecological interactions among habitat patches influence the persistence and distribution of species across forested landscapes.

In Part II we explore the dynamic ecological relationships and processes that sustain, and sometimes threaten, biological diversity. We begin by describing in Chapter 4 patterns of biological diversity and the ecological dynamics generated by recurrent disturbances within forests. These dynamics occur at a variety of scales, generating a heterogeneous mosaic of patches that sustains most species as a quilt of subpopulations. Persistence of particular species depends on both the availability of appropriate patches within this quilt and the juxtaposition of patches within the landscape fabric. As logging replaced many historical patterns of disturbance across the landscape, forests have changed greatly in composition, altering ecological relationships. In Chapter 5 we consider how interactions between adjacent habitat patches can affect sensitive elements of diversity. Such edge effects are now thought to represent significant threats to diversity in many temperate forests.

Many species may be further theatened as patches of appropriate habitat shrink and become more isolated from one another. We address issues of forest fragmentation in Chapter 6 and evaluate how reduced habitat area and isolation may threaten diversity. Many of these mechanisms are subtle and indirect, or require some time for their effects to be felt, forcing us to rely more on predictive models. To assess the validity of model predictions and to gain further insights into the mechanisms of species loss, forest managers need more information and reliable sources of data. In Chapter 7 we review the motivations for conducting thorough biological inventories of forests and monitoring change via ecological indicators. Carefully designed programs of research and monitoring will be needed to address questions regarding the impacts of forest management on sensitive elements of diversity. Although such efforts have usually been directed at just a few species, we argue that our growing knowledge now compels us to consider a broad range of ecological indicators, and that monitoring demographic performance and community processes may often be more informative than simply tracking populations of a few conspicuous species. Our recommendation to pursue a more comprehensive ecosystem approach mirrors points we make in Part III regarding the need to move from management in favor of particular species to a more synthetic concern for maintaining the patterns and processes that sustain diversity.

Whether forest specialists receive the financial and logistical support they need to complete adequate monitoring and appropriate research will depend on policy decisions within our land management agencies. The U.S. Forest Service faces particular requirements regarding the monitoring of indicator species. This agency also fields a highly qualified scientific staff capable of providing informed and regular input to forest managers. Unfortunately, their advice is too rarely solicited and incorporated into on-the-ground decisions (Chapters 13 and 14). Like the Interagency Spotted Owl Scientific Committee (Murphy and Noon 1992) and the Scientific Roundtable on Biological Diversity (Crow et al. 1993), we stress the need to expand research and monitoring efforts while integrating them more closely with each other and with routine forest management.

4

Internal Affairs: Patch and Disturbance Dynamics

Rare species are more likely to go extinct than common species. This simple statement carries an important lesson for forest managers: instead of wasting efforts promoting the abundance of common species, efforts should be devoted to protect rarer elements of diversity. As forests across the continent have been logged and reduced in size and age, species restricted to old-growth forests or forest interior conditions have declined to the point where many are rare. Most of this section will be devoted to elaborating these simple statements and to discussing how best to apply the ecological understanding they express.

Much of the confusion and controversy regarding diversity issues stems from failure to agree on the importance of rarity and the scale at which diversity concerns should be addressed. We therefore begin this chapter by describing patterns of relative abundance and basic concepts of diversity. Whether a species is rare or not depends on many factors, most of which are obscure even to biologists studying those species. Nevertheless, it is clear that disturbances influence community diversity and the persistence of many species by creating patchy habitats that affect opportunities for establishment and growth.

Forests experience many different kinds of disturbance, from occasional catastrophic fires and windstorms that cover thousands of hectares to the diggings of a mole. The types and patterns of disturbance dominating temperate forests have changed drastically in our era. What will be the consequences of these changes for their biota? Can current and proposed systems of silviculture adequately provide the spectrum of disturbances needed to maintain all elements of diversity? We pursue these questions by comparing past and current patterns of disturbance and their particular effects on rare and common elements of diversity. For example, disturbances obviously differ greatly in how predictable they are historically, greatly affecting the degree to which we can expect species to have adapted to them. We conclude that it is inappropriate to consider disturbances as similar to one another, and that forest managers will need to take disturbance regimes explicitly into account if they are to manage our forests to sustain their diversity.

Patterns of Relative Abundance

In his book *The Vegetation of Wisconsin*, Curtis (1959) analyzed the distribution of ground layer plant species among 32 distinct plant communities. More species (170 of 980) occurred in only a single plant community than occurred in any other presence class (Fig. 4-1). Half the species occurred in fewer than four communities. Thus, most plant species show a high degree of fidelity to some particular set of conditions. Interestingly, species that show high fidelity to a particular community also tend to be rare within their communities (Fig. 4-2). Thus, most ground layer plant species are rare, both in the sense that they are restricted to one or a few types of community and in the sense that they are not abundant even within that community.

Patterns of relative species abundance have now been studied in enough communities to make the generalization that most species are rare (Krebs 1985, Chapter 23). While the ecological explanations for this pattern are beyond the scope of this book, its implications for conservation are clear: rare species deserve protection. Furthermore, as old-growth forest habitats have become scarce, they harbor correspondingly higher proportions of rare species (see, e.g., Table 5.6, p. 65, in Harris 1984, based on data from Bowles 1963).

Concepts of Diversity

Although simple in principle, ideas about diversity have themselves diversified to the point where definitions and measures of diversity now exist in some profusion. In fact, whole books have been written on the subject (Magurran 1988). One successful textbook of ecology (Krebs 1985) devotes two chapters and 59 pages to the topic.

When conservation biologists use the term "diversity" or "biodiversity," they are often referring to more than just species richness. Within a species, diversity also exists in terms of the different age or size classes present and the genetic diversity that exists among individuals in any sexually reproducing species. Among species, diversity may also refer to the various habitats present in a region, the various communities that occupy those habitats, or even the diverse ecological processes that help maintain species and communities. We refer to all these entities from genetic alleles to ecosystems as *elements of diversity*. Given this extended meaning for diversity, it is clearly inappropriate to simplistically equate "diversity" with any single aspect of diversity to the

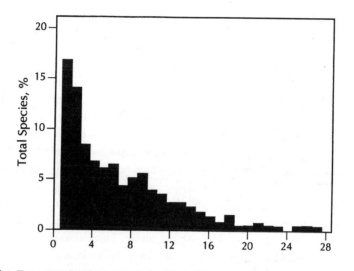

Fig. 4-1. Frequency distribution of the number of ground layer plant species occurring in various numbers of communities in Wisconsin. Note that most species occur in only one or two plant communities. [Reprinted with permission from J. T. Curtis (1959, p. 507).]

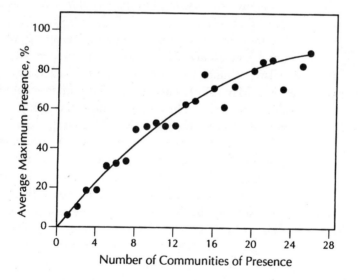

Fig. 4-2. Graph showing the relationship between fidelity (number of communities a given species occurred in) and numerical abundance (average maximum presence) for the same understory plant species in Wisconsin as shown in Fig. 4-1 (from Curtis 1959). Note that plant species restricted to one or a few communities are scarcer within the communities they occur in than species that occupy many communities.

exclusion of others. In particular, it is misleading to argue that cutting a forest stand enhances diversity because it boosts "age-class diversity" or the local diversity found within that single stand.

Ecologists conventionally distinguish multiple levels of diversity, corresponding to different spatial scales (MacArthur 1965; Whittaker 1972, 1975; Cody 1975, 1986; Pianka 1983; Magurran 1988). On the most local level, the number of species found within a particular habitat is referred to as local or *alpha diversity*. Diversity becomes augmented as we consider multiple habitats in an area. This is one basis for the increase in species numbers one encounters when sampling areas of increasing size (the species–area relationship—see Chapter 6). The change in diversity along habitat gradients is referred to as *beta diversity*. The overall diversity in a wider geographical area, then, depends both on how diverse individual habitats are and on how different the various habitats within the region are. This composite diversity across a whole region is labeled *gamma diversity*. To avoid ambiguity, whenever one refers to diversity, one should clearly indicate the spatial scale being considered. In this book, when we use the term in a general sense, we will be referring to overall regional or gamma diversity, as this is the most appropriate scale at which to base conservation efforts. In contrast, although maximizing alpha diversity is frequently pursued as a management goal, this does little to protect rarer elements of diversity restricted to special habitats or areas dominated by unique disturbance regimes.

Ecologists also discriminate between the total number of species present, or species richness, and how evenly abundances are distributed over those species. Patterns of relative abundance are the object of considerable interest within ecology (see, e.g., May 1975). This distribution is often reduced to a single number, or a diversity index. The ecological literature contains many such indices, and most depend on both species richness and relative abundance. It is of little value to choose one over another, however, since single parameter estimates generally do not distinguish among alternative patterns of expected relative abundance (May 1975). In addition, most large samples tend to assume a "canonical" log-normal distribution for reasons that may be quite general. Thus, when the entire distribution is not presented, one might as well use total species richness instead of any other more complex index.

Because resources are always limited, conservationists and foresters charged with conserving diversity should clearly concentrate their efforts regarding diversity on maintaining those particular elements most in danger of being lost or of having their ecological relationships with other elements disrupted. In addition, rare community types, rare species within communities, and species sensitive to human disturbance obviously deserve more protection than common, pervasive, or weedy species.

Disturbance and Succession

Diversity is frequently generated or maintained by recurrent disturbance. Disturbance affects diversity by determining in large part the ecological circumstances under which temperate forest species evolve. For example, disturbances often influence habitat patchiness and population demography. The context provided by a forest's characteristic disturbance regime is therefore as important as its structure in determining which species can grow there. Disturbance is customarily defined as an action that kills or removes significant biomass from a plant or plant community (Grime 1979). Disturbances therefore open patches of habitat, creating opportunities for plants and animals to recolonize.

Early American ecologists such as Cowles and Clements considered disturbance and recovery from disturbance to be central to the study of plant communities. They emphasized the often predictable way plant species tend to succeed one another following a major disturbance, returning the community eventually to a steady state. Their studies of ecological succession greatly influenced American plant ecology by injecting the notion that ecological succession is an orderly process that follows convergent paths to restore a predictable equilibrium. Their notions of recovery and convergence on a "climax" community have burdened ecology with the persistent idea that succession will automatically heal a forest of the scars of the initiating disturbance (Colinvaux 1986).

Ecologists no longer assume that plant and animal communities will automatically respond to different disturbances via the same successional sequence, or that such successions will automatically converge on a common climax community (Botkin 1990). Indeed, for years field studies have shown that succession is not necessarily predictable or convergent (Ahlgren 1974; Heinselman 1973; Connell and Slatyer 1977). The time of year a field is abandoned, how it was tilled or pastured before abandonment, its proximity to seed sources in existing forest stands, and other biological factors can all affect how succession proceeds and the composition of the resulting forest (Horn 1981). "Positive feedback switches" (Wilson and Agnew 1992), "alternative stable states" (Horn 1981), and "non-equilibrium dynamics" (Botkin 1990) have become major topics for discussion. Nevertheless, older texts and tradition continue to perpetuate the inappropriate and misleading healing analogy (see Chapter 8). Because the external environment itself changes constantly, it may also be preferable to avoid the notion that communities eventually return to any static undisturbed state. O'Neill et al. (1986) propose instead that succession within an ecological community involves returning to its predisturbance *trajectory* of change, a process they term "homeorhesis."

Without further disturbance, areas often become dominated by the most persistent and competitive species. Such species are often able to survive on low levels of resources or invest heavily in tactics to enhance their competitiveness. Typically, forest tree species that cast the deepest shade and best tolerate shade themselves come to dominate the canopy in the absence of disturbance (Horn 1971, 1975). Not surprisingly, such species are usually not as fast growing or productive as colonizing species. Thus, productivity declines in later succession and recurrent disturbance often promotes diversity by preventing dominance by one or a few species (Connell 1978; Denslow 1985).

As species disperse, establish themselves, grow, and reproduce, they do so within habitats largely determined by the dominant trees that compose the forest and by the dominant modes of disturbance affecting the community. Forests often reflect the particular fire, windstorm, or other disturbance that gave rise to them. An even deeper connection is evident in the adaptations forest plants and animals have to forest disturbances. Many forest species depend on disturbances of one kind or another to initiate or complete their life cycle. For similar reasons, many forest species are poor at tolerating novel disturbances or significant changes in the type, frequency, or intensity of disturbance to which they are adapted. Thus, anyone concerned with forest diversity must be concerned not only with forest structure but also with current and historical forest dynamics, including the scale at which these dynamics occur.

The Concept of Minimum Dynamic Area

To maintain patterns of disturbance and habitat patches similar to those that have occurred historically requires enough land to sustain disturbances up to the maximum size likely in a region and to maintain all stages of succession somewhere within the area. The area needed to sustain disturbance regimes typical of the region without losing species has been labeled the *Minimum Dynamic Area* by Pickett and Thompson (1978). This idea is central to this chapter and to all similar discussions of scale and disturbance. Species occurring in areas smaller than this are vulnerable because they may not be large enough to allow "rescue from within" (Janzen 1983) by retaining internal sources for recolonization following a large disturbance. Smaller areas might also run the risk of not being disturbed at all, thereby losing species dependent on unique early or intermediate successional stages.

As a corollary to the concept of Minimum Dynamic Area, natural areas intended to maintain biodiversity must be large and heterogeneous enough to contain multiple disturbances and all intermediate stages of recovery. Only a Minimum Dynamic Area will retain within itself viable populations of all species needed to recolonize continually disturbed sites. While it is tempting to assume that colonization from other nearby areas could occur, many forest interior and old-growth species have rather limited powers of dispersal. The Minimum Dynamic Area is therefore a function of the frequency and maximum scale of disturbance as well as the recovery potential of the most extirpation-prone species.

Types and Roles of Disturbance

Large-scale disturbances have such major effects on plant communities that their ecological importance is obvious. Hurricanes in the East and major windstorms in the Midwest often flatten extensive areas. In northerly climates, ice storms frequently damage or topple trees over a broad area. Major fires often spread for miles, resetting the successional clock. Landslides and avalanches in steep terrain often limit stand ages and impose on the landscape a mosaic of stands of varying age. Along rivers, trees tip over in the soft mud or are battered by flooding and ice.

Disturbance is so pervasive in most habitats that many forest species depend on the conditions created by disturbances to initiate or complete their life cycle. Thus, the small wind-borne seeds of many floodplain trees allow quick colonization on freshly deposited banks of silt. Similarly, the seedlings of white pine and many other species benefit from exposed mineral soils of the type that often occur after fires. Frelich (1992) concluded from his analysis of historical records that white pine occurs most frequently on sites that experience major fires at intervals of 150 to 300 years and light surface fires every 20 to 40 years. These examples illustrate how finely tuned many species are to particular patterns of disturbance. As a consequence, such species can decline markedly once these patterns change.

By opening areas for seedling establishment and colonization by other organisms, disturbances of the right kind provide key resources that promote the regeneration and persistence of many species. Within forests, canopy gaps of the size of one to a few tree crowns play crucial roles in the life cycles of many forest species (Runkle 1985). Many late successional forest tree species, such as beech and hemlock, for example, depend on gaps created in the canopy by treefalls to stimulate height growth and reach the canopy (Loucks 1970; Swain 1973; Runkle 1982). Hemlock, in fact, often experiences multiple episodes of release and suppression (Henry and Swan 1974). In addition, gap size affects which species invade. Black locust invaded larger 2-hectare openings in southern Appalachian forests to a much greater degree than openings smaller than a hectare (Phillips and Shure 1990). Thus, small selective cuts in northern hardwood forests tend to favor late successional trees like American beech, sugar maple, and eastern hemlock, while larger cuts favor yellow birch and tuliptree (Runkle 1985).

Seedling recruitment in many forest trees appears to depend on tree falls or other small-scale disturbances that commonly occur in old-growth stands. Seedlings of several dominant tree species, including hemlock, yellow birch, red spruce, and white pine, are often found concentrated on rotting logs, old stumps, or other coarse woody debris because such substrates hold moisture like a sponge and prevent desiccation (Fowells 1965; Curtis 1959; Runkle 1985, 1991). In documenting reductions in herbaceous community cover and diversity between old-growth and secondary forests in the southern Appalachians, Duffy and Meier (1992) hypothesized that many of these understory plants also depend on the pit and mound disturbances created by tree falls.

Disturbances created by animals play similarly important ecological roles in the life histories of many plants. The digging and caching of seeds and nuts by squirrels and mice serve to disperse the seeds of many trees and herbs and may improve opportunities for germination and establishment. The activities of moles and rodents often expose mineral soil and so benefit plant species that depend on mineral soil for successful germination. The brown soils of deciduous forests are in large part the result of disturbances by earthworms that, by churning the soil, mix organic and inorganic matter, greatly affecting soil texture and porosity and influencing overall rates of decomposition. In more northerly coniferous forests, earthworms are absent and phenolics and humic acids from the needle litter inhibit decomposition and leach soils of many nutrients, leaving ash-gray podzols instead. Even the micro-disturbances and patchiness created by ants are key elements in the regeneration niche for many plant species.

In the absence of treefall gaps and the other small-scale disturbances that regularly occur in mature forests, many species, particularly those common in older growth, will not encounter suitable micro-sites for establishment and releases from light competition. Without gaps and the opportunities they provide, some species will decline in abundance, affecting forest composition (Brewer 1980). Scientists who model forest dynamics realize this and have for many years based detailed simulations of forest growth on such gap dynamics (Botkin et al. 1972; Shugart and West 1977, 1980). Indeed, recurrent small-scale disturbances are central to forest dynamics and are an integral element of the steady-state conditions these dynamics may achieve (Bormann and Likens 1979a). The "climax" itself thus reflects a dynamic equilibrium of characteristic small- and medium-scale disturbances rather than a static state.

When the chance that any given type, size, and intensity of disturbance remains constant for a reasonable period of time (i.e., for several times the interval between rare major disturbances), a characteristic distribution of patches will arise and stay more or less constant. In other words, the landscape will reach a steady or stationary state where, at any given time, different individual patches are being disturbed or are recovering from disturbance, but the distribution of patch types, sizes, and ages stays the same. In fact, ecologists interested in the history of prior disturbances in temperate forests typically study this distribution of patch types and ages rather than the disturbances directly (Lorimer 1980). In other cases, this dynamic equilibrium may not occur. Theoretical models suggest that when disturbances are large relative to stand size or when disturbances occur frequently relative to the time it takes to recover from disturbance, the landscape may undergo continuous change without converging on any dynamic equilibrium (Turner et al. 1993).

The susceptibility of any particular species or community to disturbance depends on both the nature of the disturbance and the characteristics of the species or community. Many species found in frequently disturbed habitats are adapted to those disturbances. Some even appear to foster the type of disturbance they are adapted to, as in resinous pitch pine stands or chaparral, where accumulations of flammable litter make fires ever more likely. Disturbances also interact with one another in complex ways. For example, as ponderosa pine stands in the western mountains age, they become increasingly susceptible to infestations by bark beetles, who themselves carry a disease. Older stands are also, however, more likely to burn, reducing populations of the bark beetle and consequent disease transmission (Muir and Lotan 1985). Because rates of disease transmission can greatly influence the evolution of disease virulence (Ewald 1988), prevailing disturbance regimes and the resulting patchy structure of plant and animal communities can have profound impacts on the interactions of diseases with their hosts. Such complex ecological and evolutionary interactions ultimately greatly affect the relative abundance of species and the distribution of communities across the landscape.

As particular types and intensities of disturbance become common, those species whose ecological characteristics adapt them to those disturbances will increase in frequency and abundance, while those unable to tolerate or take advantage of those disturbances will decline. As the species adapted to the new disturbance regime increase in abundance, their success builds on itself as these species are better able to saturate local "safe sites" for establishment (Horn 1981). This implies that species reductions due to a historical shift in disturbance regimes (and invasion of weedy species adapted to the new disturbance regime) may persist even after disturbances return to their historical patterns. Again, we see that history matters and that effects may accumulate through time.

Thus, the size and frequency of the disturbances typical of a region greatly influence the character of its biota and do so in ways that may be difficult to reverse. Pervasive post-settlement changes in the pattern of large-scale disturbance are having profound and lasting effects on many species and their ecological relationships. Similarly, the nature and frequency of small-scale disturbances in our forests have radically shifted in our highly managed and often even-aged stands used for commercial timber production. Such forests now more rarely experience the regular, small-scale disruptions of individual treefalls and the ensuing ecological processes that so enrich mature forests. The ecological importance of these aspects of disturbance suggest that progressive foresters should attempt to mimic natural disturbance regimes when designing systems of silviculture intended to perpetuate native diversity (Runkle 1991; Hansen et al. 1991; Crow et al. 1993).

Describing Disturbances

Because disturbances are not equivalent, we should always seek to be as specific as possible when discussing ecological disturbance. At a minimum, those referring to forest "disturbance" should identify the type of disturbance being discussed and describe its size and typical frequency. Although it may be difficult to ascertain all the following characteristics for a given region, disturbance regimes are too ecologically important to characterize weakly or incompletely. Indeed, our growing understanding of the importance of disturbances in the life history and population dynamics of many species should motivate us to seek further information of this kind, especially when policy decisions are being made regarding changes in disturbance.

1. *Type.* Different disturbances affect habitats and ecological communities in different ways. How an agent removes plants or animals from an area clearly affects what propagules are left behind, how new ones colonize the site, and the prospects of both for continued growth. In addition, disturbances of different types typically differ in scale, frequency, and intensity, also inducing different effects on forests. In response, many species have adapted in diverse and unique ways to distinct types of disturbances. For example, fire-adapted species often have corky bark or serotinous cones that only open when heated. Such traits are clearly unrelated to these species' ability to respond to windthrow or commercial logging.

   Many different types of disturbance affect biological communities (see White and Pickett, 1985, for an epic catalog). Here we discuss two with important impacts on temperate forests.

(a) *Fire.* Within fire-adapted communities such as chaparral or prairie, fires recur regularly enough that most species present are adapted to tolerate the fire (e.g., by placing growing points and storage organs below ground) or recolonize quickly after fires. Many temperate forests are subject to fires on some interval. Lowland and mesic old-growth stands probably burned only rarely, while fire appears to have been a major disturbance in many coniferous stands (Ahlgren 1974; Heinselman 1973, 1981; Runkle 1985). Pine stands on sandy, nutrient poor soils probably burned most often. Traits typical of fire-adapted plant species include wind-borne seeds (e.g., fireweed); seed dormancy, as with pin cherry (Marks 1974); shade intolerance (e.g., many pines); relatively rapid growth, often adapted to the typical period between fires; a juvenile stage resistant to fires (e.g., the "grass" stage of longleaf pine); thick, corky bark and short branches or twigs (e.g., bur oak); and serotinous cones (e.g., pitch pine) or fire-stimulated germination.

(b) *Wind.* Wind disturbs forest structure on almost every scale, from dislodging twigs and branches through individual treefalls to massive blow downs. Hurricanes exert great influence on the forests of Florida and the East and Gulf Coasts, and these effects often persist for decades or even centuries (Chabrek and Palmisano 1973; Spurr 1956). In the Midwest, tornadoes and other windstorms accompanying thunderstorms often have similar effects, although usually on a more local scale (Henry and Swan 1974; Bormann and Likens 1979a). Zones of wind damage may also precipitate subsequent waves of further damage and regeneration (Sprugel 1976).

2. *Size.* Disturbances occur in all sizes, from minute (e.g., footprints that can influence seed germination in some grasses) to global (e.g., the asteroid impact that is assumed to have ended the Age of Dinosaurs at the end of the Cretaceous). In forests, larger-scale disturbances such as widespread fire or massive windstorms occur less frequently and often less predictably than smaller-scale disturbances such as treefalls.

3. *Frequency.* How often a particular type of disturbance occurs strongly affects its ecological impact and how species respond to it. Floods, for example, frequently occur on an annual basis. In a fire system such as prairie or chaparral, fires may recur fairly predictably every 2 to 20 years. Large-scale disturbances such as big forest fires and windstorms, however, usually only recur at a frequency of once every few centuries.

Disturbances that occur quite rarely relative to the life spans of the organisms that make up a community are sometimes referred to as "catastrophes" (Harper 1977). In evolutionary terms, we can only expect species to adapt to disturbances that have occurred regularly enough to have consistently favored particular traits.

4. *Intensity.* Disturbances of the same type often occur at different intensities. A light ground fire is different in character and has far different ecological effects than an intense slash or crown fire. Similarly, windstorms may be intense or mild and floods may be gentle or torrential. Such variation in intensity produces corresponding variation in impacts (although impacts obviously depend on the species affected—Menges and Waller 1983). White and Pickett (1985) distinguish "intensity" from "severity" based on whether a disturbance is measured relative to its physical force or its biological impacts, respectively.

5. *Variance and predictability.* To fully characterize a disturbance, we should also include information on the dispersion, or variation, that typically occurs in its size, frequency, and intensity. Because no two disturbances are exactly alike, variation among disturbances will greatly affect how species respond to both individual disturbances and general classes of disturbance. Some disturbances occur quite regularly and predictably, while others occur only occasionally or randomly. Regular disturbances may result either from the regularity of recurrence of external disturbance events or from a pattern of increasing susceptibility or sensitivity to randomly generated disturbances as time since disturbance increases. Infrequent, large-scale disturbances are often less predictable and so may represent catastrophes to the species present.

In general, the more regular and predictable a disturbance has been over evolutionary time, the more species will have adapted to it. Conversely, few species can be expected to thrive or take advantage of novel or unpredictable disturbances. For example, the regular tides that have washed ocean margins for eons support a biotically rich and highly productive marine community in the intertidal zone, including clams, mussels, barnacles, many algae, and the animals that prey on these. In stark contrast, seemingly similar zones where freshwater levels fluctuate significantly once or twice a day, such as the edges of rivers below hydroelectric dams or lakes now used for pumped electric power storage, support very low cover and diversity. Thus, there is an evolutionary basis for adaptation to disturbances that cannot be ignored, and distinct types of disturbances cannot be considered analogous in terms of their impacts on disturbance or community recovery.

Characterizing Disturbances

Given the ecological importance of disturbance regimes and the need to distinguish differences among them, we need tools to characterize disturbance regimes in a succinct fashion. Because disturbances differ in so many ways, however, this is not a trivial task. Most simply, one can compare the sizes (or any other single characteristic) of a particular type of disturbance by graphing them in a frequency histogram (e.g., Fig. 4-3). Comparing two such histograms directly indicates how they differ in the average size (or other character) of disturbance. These distributions also amply illustrate the range, or variance, in that type of disturbance at a site. Of course, it is always possible that the two areas differ in some other aspect of their disturbance regimes, making, in principle, several such histograms necessary before one could conclude that the two areas did not differ in their disturbance regimes.

To capture two significant elements of the disturbance regime at once, one can also plot the frequency of disturbances in a phase plane where each axis represents a different aspect of the disturbance (e.g., size and frequency—Fig. 4-4). Such *phase diagrams*, introduced by Delcourt et al. (1983), summarize considerable information and facilitate comparisons of regions or periods whose disturbance regimes may differ in two or more respects. We can refer to the distribution of sizes, frequencies, and intensities of a given disturbance as the *disturbance spectrum* for that region. While disturbance spectra remain incompletely described for many forest types and regions, they represent a fundamental characteristic for all forest communities. They are also assuming increased importance in debates regarding the ecological appropriateness of

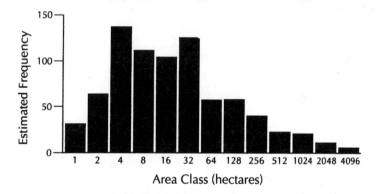

Fig. 4-3. Frequency histogram of the distribution of windthrow sizes in mid-nineteenth century forests in northern Wisconsin. The data were culled from General Land Office survey records by Canham (1978) and may significantly underestimate the number of small-scale disturbances. Note that area classes are divided into a log series, suggesting that this type of disturbance is distributed approximately log-normally.

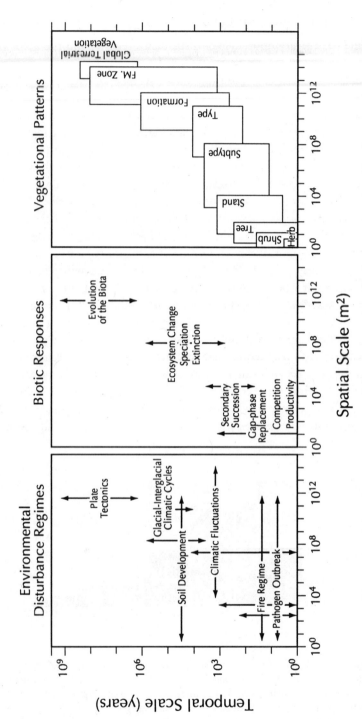

Fig. 4-4. The broad scales of space and time that pertain to environmental disturbance regimes, biotic responses to these disturbances, and patterns of vegetation. Note that disturbances affect temperate forests over a wide set of spatial scales (x-axis) and intervals between disturbances (y-axis). [Reprinted with permission from Delcourt et al. (1983). See also Delcourt and Delcourt 1991 and 1992.]

various alternative systems of forest management (Chapter 9). We can use phase diagrams and disturbance spectra to facilitate comparisons between historical disturbance regimes (e.g., Fig. 4-5). Such comparisons should further include differences that typically exist in the intensity of disturbances (Fig. 4-6). To make these comparisons more concrete, we now turn to what is known regarding typical patterns of disturbance in temperate forests.

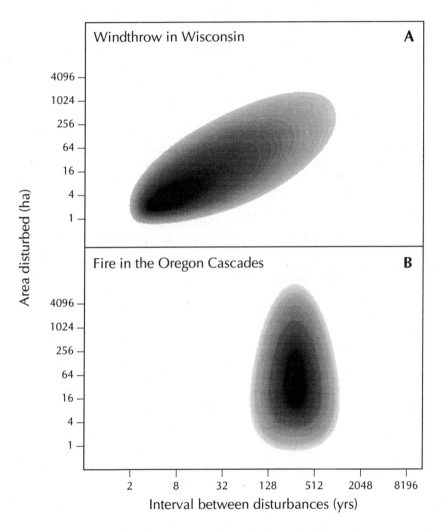

Fig. 4-5. Hypothetical disturbance spectra for predominant disturbances in two temperate forest areas. (A) Windthrow disturbances, based roughly on the information presented by Canham and Loucks (1984). (B) Fire disturbances in the Oregon Cascades, based on information provided by Morrison and Swanson (1990).

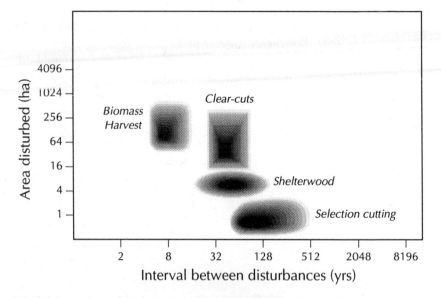

Fig. 4-6. Approximate disturbance spectra for different modes of timber harvest, including selective cutting, shelterwood harvest, clearcuts, and harvests for biomass production. Note differences from historical disturbance sizes and frequencies in Fig. 4-7.

Historical Patterns of Disturbance in Temperate Forests

To understand how the disturbance regimes typical of North American temperate forests before European settlement compare with those that dominate today, we need an accurate picture of historical disturbance regimes. Fortunately, ecologists and paleobotanists have devoted some effort to this question and have devised a number of methods that allow us to infer these disturbance regimes with some certainty. In this section, we review what is known about the characteristics of disturbances that have historically influenced forest structure and dynamics.

Fire, wind, and perhaps ice storms have all had impacts on the structure and processes of temperate forests (Runkle 1985). Their relative significance, however, and the degree to which they tend to dominate forest structure and succession vary considerably. For example, within the hemlock-hardwoods region of the upper Midwest, several types of disturbance play roles in maintaining a regional mosaic of patches and corresponding diversity. As in other mature forest types, treefall gaps occur commonly and represent a significant part of the dynamic equilibrium that maintained this community type (Curtis 1959; Runkle 1985). Rarer, large-scale disturbances, including fires and windthrow, opened up larger areas, making them more suitable for invasion by herbs, shrubs, and light-demanding tree species. Studies of presettlement forests in Wisconsin suggest that at any one time 17 to 25% of Wisconsin's northern woods were early- to mid-successional stands recovering from large-scale windthrows (Canham and Loucks 1984).

A disturbance that occurs frequently and predictably enough to affect most of the species present in a community is likely to exert a dominant effect on that community. For example, fires dominate as a recurrent disturbance within pine stands in many barrens habitats and particularly within the "pine plains" of the New Jersey pine barrens (Givnish 1981). Under such dominant disturbance regimes, most individuals within a stand will be even-aged, and the overall forest will exist as a mosaic of discrete, differently aged patches. If, instead, disturbances of various kinds occur and no single type occurs frequently or extensively enough to dominate a region, stands of diverse age will occur interspersed throughout the forest.

Both dominant and mixed disturbance regimes appear common in temperate forests. Fires have apparently been a dominant mode of large-scale forest disturbance in the mixed hardwoods of northern Minnesota for the last 10 millennia (Swain 1973). Working in the Boundary Waters of Minnesota, Heinselman (1973, 1981) noted that essentially all stands date back to a fire, but different stands were initiated by different fires. The probability of a fire in any given year was about 1%, meaning that few stands would go longer than a few hundred years without experiencing a major fire.

Fire and large-scale windthrow together have dominated white pine stands on sandy flats and terraces on the Allegheny Plateau (Hough 1936; Runkle 1985). In mesic upland sites in the same region, eastern hemlock and American beech dominated mature stands. Here, fires were rare or absent due to cooler, moister conditions, allowing sporadic individual treefalls and other small- to medium-sized gaps to dominate forest dynamics (Hough and Forbes 1943). In these old-growth stands, only about 0.5% of the area was disturbed in any given year.

Runkle (1982, 1985) found that similar small-scale disturbances predominated in southern Appalachian cove forests. The distribution of gap areas in his study followed a log normal distribution (that is, a normal distribution on a logarithmic scale), with about 1% of the land area in gaps of 400 square meters or more at any given time. He estimated that from 0.5 to 2% of the forest was disturbed by new tree- or limbfall gaps in any given year, and that the trees surrounding gaps were therefore often likely to encounter multiple gaps during their lifetimes.

Lorimer (1977) estimates that the interval between major fires in northeastern Maine was at least 800 years. These shady old-growth forests harbored considerable moist rotting material and had few low limbs to pass fires to the crown. Inter-fire intervals were somewhat shorter in the pine forests of southern Maine, averaging 530 years (Fahey and Reiners 1981). Although hurricanes batter New England on a regular basis, Lorimer estimated the interval between such major windthrow at any given spot in the region to be at least 1000 years. Outbreaks of spruce budworm were also apparently rather rare, in part because only a small proportion of the forest was made up of the susceptible balsam fir.

Runkle (1985) notes that most temperate tree species reach full maturity by the time they reach 300 years. He proposes that this age corresponds to both the increased susceptibility of larger and older tree crowns to disturbance and to the ensuing typical time between disturbances in these communities. Indeed, evolutionary considerations suggest why this might be the case. We should expect species' longevity, in general, to be adapted to the average interval between disturbances, as species with much shorter life spans will generally be edged out by longer-living species, while species with much longer life spans will typically encounter a disturbance before they reach reproductive maturity. Such adaptations by many of the species within a community also account for the observation that species diversity typically reaches a maximum when disturbances occur at a historically intermediate rate (Connell 1978; Denslow 1985). This "intermediate disturbance hypothesis" has obvious implications for forest management.

How Do Present Patterns of Disturbance Differ from Historical Patterns?

Some foresters consider silviculture inherently similar to other types of ecological disturbance and have sought to justify timber harvest activities on the grounds that they are analogous to natural disturbances. Small-scale selective cuts are considered similar to individual or patch treefalls from natural disturbances such as blowdowns, while clearcuts often are compared to large-scale blowdowns or fire. Recovery following logging operations has therefore been considered a typical type of secondary succession in which the forest restores itself to its former state. Given this assumption, logging should pose no threat to retaining forest species, as all should be adapted to disturbance.

Timber industry spokesmen in Maine promoted the idea that forests in northern New England were naturally even-aged due to frequent large-scale fires, blowdowns, insect outbreaks, and disease. Lansky (1992) criticized these myths of industrial forestry by pointing out that these presettlement forests were, in fact, uneven-aged (Lorimer 1977). There also is evidence that fires and epidemics actually increased in severity following settlement.

The natural succession analogy does hold up to a point. Clearcuts and thinnings open up the canopy, exposing the ground layer to more sun and encouraging the growth of shade-intolerant species in a manner similar to fires or blowdowns. Beyond this, however, the analogy breaks down. The intensive forms of silviculture practiced today differ significantly from the disturbances that once ruled the landscapes they cover. For example, both fires and windstorms disturb forests in patches, sparing many trees and pockets of habitat that serve as inocula to recolonize surrounding areas. Fires also release seeds from serotinous cones or break dormancy in the seeds of fire-adapted species. In addition, they release a pulse of nutrients used by many of the species colonizing the site and often warm the seed bed by blackening the earth. Windstorms not followed by fire leave a jumble of coarse and fine woody debris that provides nutrients, some shade, and protection from browsing ungulates.

Silvicultural treatments like thinning and selective cutting alter the scale, frequency, and intensity at which forests are disturbed (Figs. 4-6 and 4-7). In northern Wisconsin, managed forest stands are mostly either even-aged stands of early successional species such as aspen, red pine, or paper birch or uneven-aged stands of mixed hardwoods. In either case, younger stands are less susceptible to windthrow than old growth, reducing the frequency of small- and large-scale blowdowns. Even-aged stands also are less likely to contain variably sized gaps due to individual tree blowdowns. Such stands cannot be expected to generate as many gaps and tip-up mounds or as much coarse woody debris as mature stands, possibly restricting opportunities to regenerate the late successional species that depend on these micro-sites. Such effects may account, in part, for the reductions in seedling recruitment noted for species such as eastern hemlock and white pine.

In addition to changing the scale and intensity of disturbance, logging activities have also greatly altered the incidence of natural disturbances. Most conspicuously, the incidence of large-scale fires and windthrow has been reduced. This reflects both policies of fire suppression and the extensive conversion of mature forest to younger stands less susceptible to windthrow (Canham and Loucks 1984). Most fires in the northern and eastern U.S. are now extinguished as soon as possible to limit threats to commercial timber stands and human settlements. Dense networks of roads and fire lanes facilitate this policy and often act as fire breaks to limit the propagation of fires. Fire suppression has also threatened some fire-adapted communities that depend on fire to exclude invasion by certain woody species. By limiting ground fires that consume fuel, fire suppression also tends to increase fuel loads, magnifying the likelihood of intense crown fires in drier forests. As the effects of fire vary in different contexts and the general public perceives fires as destructive, it is hardly surprising that fire policies on public lands remain controversial (Pyne 1982, 1989).

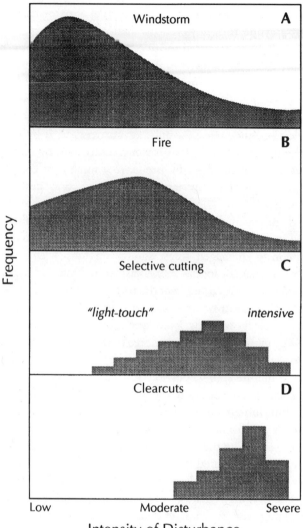

Fig. 4-7. Comparison of the intensities of forest disturbance between historical disturbance regimes involving windthrow (A) and fire (B), and contemporary silvicultural disturbances from selective harvests (C) and clearcuts (D). Selective harvests vary considerably in intensity, depending on the proportion of trees cut and the type of equipment used. Clearcuts here are assumed to involve extensive road building, the use of heavy equipment, and subsequent site preparation (disking or herbicide application).

Norse (1989) summarized many of the ways managed forests in the Pacific Northwest differ from primary forests. Many of these distinctions extend to other temperate (and probably tropical) forests. For example, in landscapes managed for commercial timber production, logging is normally scheduled to maximize economic returns and so generally occurs when tree growth rates start to slow (the culmination of mean annual increment). This occurs in what an ecologist might term "middle age." Maximum ages are far greater, but trees are rarely allowed to mature to such age (or size) in "well-managed" stands. Traditional foresters have often considered standing or fallen dead trees as wasteful. In seeking to minimize their occurrence, they also reduced their ability to play crucial ecological roles as roosts, habitat for cavity nesting species, substrates for fungi and invertebrates, etc. The tip-up mounds and logs created by windthrow have also been considered a nuisance rather than critical micro-sites for regeneration. Today's clearcuts occur more frequently and remove more above-ground biomass and nutrients than the natural disturbances they replace. In addition, modern heavy logging equipment exposes and compacts the soil more than natural disturbances do, contributing to erosion and the loss of soil nutrients.

Perhaps as serious as any of these other changes are the shifts in relative abundance that have occurred in managed landscapes. As early successional species have become much more abundant, they have come to dominate the habitats they occupy and the seed pools available to colonize suitable micro-sites that appear. As noted above, such profound shifts in abundance can re-direct community change in entirely new directions. In addition, their increased abundance has been used to justify continued management in their favor, perpetuating short cutting cycles and the altered disturbance regimes they represent.

Thus, conventional silviculture greatly modifies dominant modes of disturbance and constrains the degree to which species dependent on traditional disturbances can prosper in altered landscapes. It remains to be seen to what extent we can re-design silvicultural systems to accurately mimic the natural disturbances that play such an important role in many species' lives (Chapter 10).

Patchy Resources and Populations

To maintain the biodiversity of a region, we need to maintain at least minimally viable populations of all species present. Population viability, in turn, often hinges on population size: the more individuals present, the less likely that particular population will go extinct. Demographically minded biologists frequently go further to assess viability in terms of how a population is made up of individuals of various sizes or ages. More recently, ecologists have begun to pay more attention to how populations are distributed spatially and how distinct subgroups interact to determine the population viability over a whole region. More often than not, patterns of distribution appear to reflect dynamic processes like colonization in relation to disturbance.

Most wild populations are more or less discontinuous, occupying a mosaic of patches. At any given time, such populations are actually composed of individuals distributed among patches of appropriate habitat, with each patch composing a more or less distinct subpopulation that is relatively independent of the others. Individuals within any given subpopulation tend to primarily interact and mate with other individuals within the same subpopulation. At any given time, some patches contain subpopulations that are viable and growing, while other patches support only a few scattered individuals or a declining subpopulation. Although relatively distinct, individuals occasionally move among these subpopulations, allowing some genetic exchange and the periodic "rescue" of declining or extinct subpopulations. Within such a *metapopulation,* there may be continuous turnover: new subpopulations are founded while some existing populations continually wink out of existence. In such situations, the long-term persistence of the metapopulation in the region clearly depends on the balance that exists between growing and declining subpopulations. In addition, some patches remain unoccupied if individuals from other subpopulations are too far away, too few, or otherwise unable to colonize that patch.

As long as new subpopulations are founded at least as fast as established subpopulations disappear, the species will persist in a region. Thus, metapopulation dynamics have become a central focus of research within conservation biology (Gilpin and Hanski 1991). Many are now investigating the factors that cause some subpopulations to consistently act as *sources,* sending out a surplus of individuals to colonize other areas, while other subpopulations act as *sinks* by sustaining only population growth at a rate too low to sustain its presence. To answer such questions, many scientists are adopting landscape approaches that explicitly encompass spatial relationships such as the proximity of subpopulations and their relationship to the distribution of disturbances. Scientists have also become interested in learning how changes in the dominant disturbance regimes influence population dynamics and persistence by changing the spatial distribution of habitat patches and the corresponding distribution and performance of subpopulations.

Changes in the predominant disturbance regimes in a region are likely to have effects on many elements of diversity. When disturbances of a certain type become more or less severe, more or less frequent, or change significantly in some other respect, some species will benefit from the change while other species will suffer. Furthermore, because most species will generally be adapted to those disturbance regimes that have historically dominated an area (Denslow 1985), any significant departure from these regimes is likely to eventually precipitate a net reduction in overall diversity. Thus, *a priori,* one can expect *any* significant alteration of historical disturbance regimes to cause a net decrease in overall diversity. This may not be apparent immediately, however, as new species that benefit from the change in disturbances may appear quite quickly, while the species adapted to historical conditions may decline rather slowly as metapopulations become thinner and sink populations begin to outnumber source populations.

The Importance of Scale

Because disturbances and patchiness occur across a wide range of scales, any attempts to retain natural processes of disturbance and associated patch sizes must take scale into account (Chapter 6). Small-scale disturbances such as treefalls are part of the internal dynamics of almost any sized mature forest. Larger-scale disturbances like fires or blowdowns, however, obliterate entire forest stands, as occurred on July 4, 1977, when a downburst storm flattened the "Big Block" in the Flambeau River State Forest. This State Natural Area was one of the last virgin forests left in Wisconsin, demonstrating the vulnerability of small stands to individual disturbances.

In general, landscapes not significantly larger than the size of typical large-scale disturbances will be affected to a far greater degree by such disturbances, producing large fluctuations in their long-term ecological dynamics (Turner et al. 1993). If the disturbance frequency also is increased relative to the recovery interval, there may even be unstable or catastrophic change. Turner et al. concluded that "the disturbance could fundamentally change the nature of the system if the species cannot become reestablished." These theoretical results alert us to the danger that smaller areas may be vulnerable to irreversible changes in ecological structure and processes, particularly when reductions in area are coupled to changes in the disturbance regime.

Management Implications

If we consider it important to maintain all forest tree species and successional types that were once found in an area, we face a choice. Most simply, we could reserve from active management sufficient contiguous lands to provide for most scales of natural disturbance, passively promoting the reestablishment of historical disturbance regimes. These areas intended primarily to conserve diversity would detract from the acreage devoted to active silviculture, as they would need to be rather large to be efficient and effective (see Chapter 11 for a specific calculation). This approach is conservative in the sense that it requires minimal information—namely, the typical maximum size of disturbances. It also requires very little active management, provided natural disturbances (including at least some fire in many regions) could be accommodated. Finally, this approach coincides with prescriptions to retain diversity based on area concerns (Chapter 6).

On the basis of the Minimum Dynamic Area concept, how large should a biotic reserve be in the upper Midwest? Canham and Loucks (1984) estimated from General Land Office survey records that although mean windthrow area is about 93 hectares, the maximum size is about 3800 hectares (Fig. 4-1). Thus, to allow a natural reserve to sustain such a disturbance and still retain no more than 25% of its landscape in recovery from this single disturbance would require designating an area of at least 15,000 hectares. Given that the time for recovery is 200 to 300 years in this forest type, it would be safer to allocate 20,000 hectares or more (see Chapter 10). It is also possible that smaller areas might suffice to retain all species if they were relatively near each other and connected via a biologically effective network of corridors (see Chapter 6).

Alternatively, forest managers might attempt to conserve the historical disturbance regimes and species dependent on them by devising ways to artificially mimic natural disturbance regimes within the structure of some redesigned system of silviculture (see Chapter 10). Several recent proposals have some promise if they can succeed in ameliorating the often disrupting effects of intensive forestry on native biodiversity (G. Robinson 1988; Franklin 1989, 1992). Nevertheless, most of the proposals depend on frequent access via roads, raising issues regarding edge effects and related road impacts (Chapter 5). Moreover, they assume considerable knowledge and certainty regarding the roles of original disturbance regimes and the degree to which silvicultural manipulations can substitute for these. Until and unless we can do the research necessary to demonstrate this substitutability (Chapter 7), it will be premature to expect that these proposals will suffice to maintain all species dependent on original disturbance regimes and forest structures.

Summary

Disturbances across a landscape usually result in a complex mosaic of patches of various sizes, ages, and types. This heterogeneity contributes significantly to overall diversity and the metapopulation structure of many forest species. Patterns of disturbance greatly influence the distribution and abundance of species by affecting the nature and distribution of habitat patches and consequent opportunities for population establishment and growth. Forest stands below a threshold termed the Minimum Dynamic Area, however, cannot be expected to sustain their full range of patch sizes, ages, and types, making them vulnerable to losing elements of diversity.

The ability of species to colonize disturbed sites or invade gaps clearly depends on the nature of the disturbance, how often it is occurring on the landscape, and how close such disturbances are to preexisting populations of that species. Furthermore, whether a given habitat patch is appropriate and the likelihood that dispersing individuals will find it depend critically on the overall types, frequencies, and scales of disturbance occurring in the forest. These ideas provide a logical connection between ideas regarding disturbance and ecological notions regarding "patch dynamics" (Levin and Paine 1974; Pickett and Thompson 1978).

Although we have barely begun to understand these complex interactions, it is clear that primary, old-growth forests experience different kinds, frequencies, sizes, and intensities of disturbance than secondary managed forests. It is clearly inappropriate to simplify the ecological effects of disturbance by equating various types of disturbance with one another. Aside from their intrinsic differences in structure, younger forests support quite a different set of disturbances, further altering the ecological setting in which species interactions occur.

From Hero to Villain:
Edge Effects

Ecological processes operating within formally protected nature reserves, or within other specific blocks of habitat, shape the biota of these areas through time. Managers must be keenly aware of ecological processes operating within these areas, such as patch dynamics and population stochasticity, if they are to maintain resident biodiversity. Yet, biological phenomena originating outside the boundaries of these areas also strongly influence the fate of the resident biota. This complicates the manager's job greatly, since it implies a need for awareness and responsibility for biological events occurring in much larger portions of the landscape.

Many interactions of forested nature reserves or other habitat blocks benefit the biotic residents of these areas. Examples include incoming streams, colonizing warblers, and movements of pollen. Other external influences are much less benevolent. During the last two decades, these deleterious external influences have come to be known as *edge effects*. In this chapter we discuss both traditional and recent usages of this term and their significance for protecting forest biodiversity.

Competing Concepts of Edge and Edge Effects

An edge can be defined as the interface between two different types of habitat, such as the border between an old field and an adjacent woodlot or the border between mature and younger forest stands (Thomas et al. 1979; Forman and Godron 1986). Edges can occur naturally, or they can be produced through human activities such as agriculture, silviculture, management for game animals, and residential development.

To our knowledge, Aldo Leopold was the first to articulate the concept of edge effects (Leopold 1933). In *Game Management*, Leopold discussed game abundance as an edge effect, arguing that an increased "interspersion" of different habitat types would benefit game "of low mobility and high [habitat] type requirements" by providing simultaneous access to different habitats. In support of his theory, he noted the kinds of edges instinctively sought by hunters of grouse, quail, snipe, deer, rabbits, and ducks, suggesting it was common knowledge that a direct relationship existed between game management and the amount of edge habitat available.

The idea that managed habitats can and should be manipulated to increase the amount of edge so as to provide a bountiful crop of game soon became firmly established in the literature of wildlife ecology (Graham 1947; Trippensee 1948; Dasmann 1964). Although some of the ecological mechanisms responsible for this plethora of game need further investigation (e.g., McCaffery 1986; McCaffery and Ashbrenner, in prep.), few biologists today would contest the fact that edges, in general, are good for game species. But are they good for wildlife?

Approximately 20 years ago, biologists began to notice that the same edges that were so beneficial to populations of game species had a more varied effect on other species of wildlife (Burger 1973; Hunter 1990). In the ensuing decade, many studies were published which specifically discussed the detrimental effects to certain nongame wildlife as a direct result of the creation of edges (Gates and Gysel 1978; Ranney et al. 1981; Ambuel and Temple 1983; Brittingham and Temple 1983; Guntenspergen 1983; Janzen 1983). This literature has grown considerably during the last decade, focusing primarily on edges as perceived by birds, mammals, and vascular plants (reviewed in Reese and Ratti 1988, and Yahner 1988; see also Harris 1988a; Bellinger et al. 1989; Concannon et al. 1990; Kricher 1990; Small and Hunter 1989; Benjamin and Maguire 1991; Gates 1991; Gates and Giffen 1991; Noss 1991b; Saunders et al. 1991; Walters 1991; Chen et al. 1992; DeGraaf 1992; Howe et al. 1992; Mills 1992; Nee and May 1992; Brothers 1993).

The concept of edge effects reflected by these two decades of nongame literature differs greatly from that originally proposed by Leopold—a dark side of the same management strategy. Under this alternative view, edge effects are seen as ecological phenomena detrimental to populations of wildlife adapted to forest interior conditions. These edge effects occur when human activities, such as logging or road building, create artificial kinds and amounts of edge on the landscape. The clash between this relatively recent, negative concept of edge effect held by conservation biologists and the traditional, positive concept held by many game managers has led to an intense debate over the methods of public land management (S. K. Robinson 1988). At issue in this conflict is the degree to which we as a society promote game species over certain nongame species through the creation of edge habitats. Noss (1987a) and others have criticized the practice of creating large amounts of edge habitat across the landscape in an attempt to "maximize diversity." This practice (Fig. 5-1) is promoted by many land managers who believe that, since

Fig. 5-1. Dispersion of clearcuts in an older forest matrix tends to maximize edge habitats, favoring some wildlife species but not others. [Photo: D. Waller.]

edges maximize the diversity of many animals on a local basis, a replication of such habitats throughout a forest will maximize overall diversity.

The reasoning behind this common practice is fallacious for several reasons. From basic tenets of diversity theory (Chapter 4), we know that maximum forestwide diversity will not be achieved by endlessly replicating the most diverse individual habitats but instead by including functional representatives of other habitats, which, although sometimes less diverse themselves, may include elements of biodiversity not represented elsewhere. Too great a replication of edge habitats throughout a forest reduces diversity at forestwide and regional scales (WFCTF 1986; Noss 1987a).

Traditional reasoning errs in two other important ways regarding edge effects. First, it assumes that what is good for game is good for all wildlife within these relatively diverse edge habitats. By the confusion of game with wildlife (i.e., *all* species living in the habitat), the effects of management for edges on nongame wildlife were long overlooked (Peek 1986; Schowalter 1986; Chapter 2). In this way, traditional concepts of edge fail to encourage an accurate accounting of the actual benefit and detriment produced by artificial edges on wildlife as a whole. Traditional reasoning also errs by assuming that the benefits and detriments of management for edge habitats are relatively localized. Because the positive effects of wildlife openings on game species such as ruffed grouse are relatively local, managers attempt to scatter many of these openings across the landscape so as to boost game populations throughout the forest (e.g., USDA Forest Service 1986a). However, the negative biological effects of these openings may operate at different and sometimes greater scales.

Conceptualizing Edge Effects

The striking conceptual differences between the traditional (positive) and recent (negative) definitions of edge effects disappear if the terms "edge" and "edge effect" are used in a more precise way, that is, if conclusions are drawn only with explicit reference to the biology of individual species or specific groups of species. For example, if traditional wildlife managers had been more precise in their statements about the benefits of management for edge habitats, they would have concluded from their data only that benefits for game and some, not all, species of wildlife were produced, as did Leopold in 1933. Only when their conclusions were extended to include all wildlife, without supporting data, was a problem created in both conceptual and actual terms. Nontraditional wildlife managers and conservation biologists need to be equally careful about generalizations from our knowledge of edges and edge effects. In our thinking and writings, we sometimes draw erroneous conclusions about the degree to which one species or ecological phenomenon can stand for another. We often conceptualize edge in overly simplistic terms (Bradshaw 1992).

A substantial proportion of the early literature on negative edge effects focused on abiotic factors such as the increased penetration of sunlight and wind along the newly exposed edges of forest stands adjacent to logged areas. Many of these types of edge effects (with exceptions, below) are relatively limited in scope, penetrating newly exposed edges for only 50 meters or so. Furthermore, some of these abiotic edge effects are relatively amenable to measurement, e.g., the temperature and relative humidity within a remnant forest stand at various distances from a clearcut. Once such measurements have been taken, it is perfectly reasonable to talk about the percentage of the remnant forest stand that is edge, versus interior habitat, as defined by temperature and humidity criteria. But since these are only two of the many phenomena that organisms experience as edge effects, it is not reasonable to declare without qualification that $x\%$ of the stand is edge habitat and $1-x\%$ is interior, as is often done. By considering only single or few edge factors at a time, we often tend to conceptualize forest stands (or other blocks of habitat) as if there were a dotted line some distance within them that divided interior from edge (Figs. 5-2A,B). If we simultaneously consider the effect of a single edge factor, such as available light, on a number of species, their responses can and do vary widely. Increased sun at the edge of a newly isolated forest stand might stimulate the growth and reproduction of a certain vine but reduce the probability that particular fungi would flourish. This, too, should dissuade us from simplistic claims about what is edge and what is interior; the perception of edge and interior habitats depends on what you are. The obvious manifestation of this truism, that trout and hemlocks perceive edge differently, has occurred to most, but perhaps not the idea that not all fish, or all forest trees, perceive edge in the same way.

If we had sufficient empirical data to accurately visualize the distribution of negative edge effects as experienced by some particular species within a hectare of temperate forest over a year's time, the result would appear as a stippled drawing (Fig. 5-2C). Some areas would be dark from concentrations of black dots (representing actual occurrences of the edge effects) and other areas relatively light in color (representing habitat that was relatively edge-free). With such empirical data, we could also draw continuous lines through areas where this particular species experienced equal concentrations of negative edge effects (Fig. 5-2D), much like topographic contours, or the isobars drawn on a weather map to connect all points with equal barometric pressure. These isoplethic lines, perhaps best called isoacmes (from the Greek *iso* [same] + *akme* [edge or point]), would more accurately define the degree to

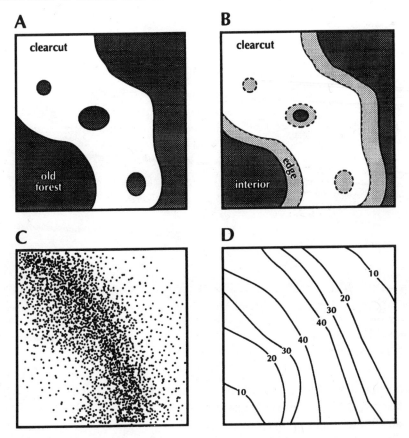

Fig. 5-2. Conceptualizing edge effects: This square block of forest is about evenly divided in area between old forest and a clearcut (A). We often conceptualize edge as if there is a distinct and real boundary between the interior of the old forest and the edge (B), but such a boundary can only be defined with reference to individual organisms. A particular species might experience negative edge events over a period of time (as indicated by black dots in Part C) in a pattern that does not correspond closely with the boundaries of habitats as we perceive them. Given data like those in Part C, we could draw "isoacmes" to delineate area of equal intensities of edge effects as perceived by this same organism (D), much like the elevational countours on a topographic map. Isoacme maps of the same landscape would differ greatly between species.

which portions of such a forest could be considered interior habitat from the perspective of forest inhabitants. Given differences in response, each species or set of similar species would require its own isoacme map.

Conspicuous, newly created edges like those formed at the boundary of a new clearcut and an adjacent old-growth stand (Fig. 5-3) have justifiably received much research attention. Fortunately, an evaluation of edge effects due to other, less apparent edges, such as those along trails and small roads

Fig. 5-3. The outer edge of an old-growth hemlock–white pine–yellow birch stand in northern Wisconsin was removed by clearcutting during the spring of 1993. Long-term monitoring plots for deer browsing within the stand are being used to measure the intensity of deer edge effects as the surrounding forest matrix changes. [Photo: W. Alverson, 1993.]

(Hickman 1990; Howe et al. 1992), and between forest stands of varying ages (Mason and Moermond 1991; Bradshaw 1992; DeGraaf 1992; Thompson et al. 1992), are now underway.

One unconventional example of an edge effect comes from the Forest Service's deliberations over the shape of three large clearcuts, 324–372 hectares in size, so as to maximize the amount of habitat temporarily provided for the sharp-tailed grouse. Because of the linear shape proposed for the 324-hectare clearcut, and the perception by sharp-tailed grouse that open areas within 200 meters of a forest edge are unsatisfactory habitat, the entirety of this clearcut would contain no interior (i.e., suitable) habitat from a sharp-tail's point of view (Kick et al. 1991; see also Rudnicky and Hunter 1993).

We are decades away from having reasonably complete information on the actual occurrence of negative edge effects in temperate (or tropical) forest stands (Lovejoy et al. 1986; Hunter 1990; Laurance 1991; Skole and Tucker 1993). In the more immediate future, the construction of isoacme maps for individual forest species of particular interest is a realistic goal. Careful field studies should be undertaken to document the responses of individual species and populations to various phenomena suspected of representing negative edge effects on a geographic basis (Yahner 1988; Bradshaw 1992), an issue closely linked to determining a more precise definition of habitat for many nongame species (Harris 1988b). Some are already underway, especially those examining vertebrates and vascular plants, but many more are needed if we are to assemble a more complete understanding of both positive and negative effects of edge habitat on wildlife.

How Deeply Do Edge Effects Penetrate Forest Stands?

Fortunately, some data exist that allow us to estimate the distances over which negative biotic and abiotic edge effects operate and the corresponding size of biological reserves necessary to contain some relatively edge-free habitat. Many abiotic edge effects influence forest stands within 50 meters of their boundaries with adjacent, open habitats. Others, such as increases in windthrow (toppling of trees), wind "fetch," and acid rain, can operate at much greater distances. Fetch, defined as the minimum distance from the edge of an isolated forest stand at which the wind profile becomes like that over a nonisolated forest stand, will often be 2–4 kilometers in magnitude (Saunders et al. 1991), resulting in considerable portions of the forest canopy of isolated stands with altered dynamics of gas exchange.

Changes in the dynamics of wind and water alter rates at which seeds, spores, insects, and bacteria are transported into (and out of) forest stands, just as changes in the abundance of these mobile biological entities in the surrounding matrix of vegetation will greatly influence the numbers introduced into remnant stands. An increase in the number and kinds of weedy species surrounding a block of forest increases both the absolute number of weed seeds dispersed into the reserve and the overall proportion of the seed rain composed of weedy species. The effect will be most immediately evident in natural or artificial gaps within forest blocks, where exotic weeds such as dandelion, burdock, selfheal, motherwort, plantains, thistles, and hybrid honeysuckles (in our region) spring from the seedbank or grow from newly arrived seeds in response to increased light levels. More subtle, cumulative changes in seedbank composition undoubtedly occur as well. The degree to which the increased presence of these weedy species within forest gaps alters other aspects of forest diversity, such as insect and fungal communities, remains largely unknown.

The buildup of animals preferring open habitats can also have profound effects on nearby, or even relatively distant, forest stands. Predators favored by edge habitats, such as raccoons and blue jays, penetrate adjacent forest stands to distances of up to several hundred meters, altering the prospects of survival for many organisms, including songbirds, herpetiles, and insects (Matthiae and Stearns 1981; Andrén et al. 1985; Wilcove 1985; Wilcove et al. 1986; Andrén and Angelstam 1988; Yahner and Scott 1988; Terborgh 1992).

Other edge-loving birds, such as cowbirds, enter forest stands and reduce the nesting success of native birds by substituting their own eggs while destroying those of the host (Brittingham and Temple 1983). It has recently been proposed that the main factor responsible for the loss of migrant songbirds from our woodlands may in fact be nest parasitism by cowbirds, a direct result of the large amounts of edge habitat created by human activities (Robinson 1992; Böhning-Gaese et al. 1993).

The creation of grassy fields, forest gaps, and young forest stands regenerating after clearcuts also boosts populations of white-tailed deer (McCaffery 1986). Because deer are highly mobile animals in our region, a population increase produced by any given clearcut may have significant consequences for forested areas at some distance away; yet the exact nature of this spatial relationship is still poorly known (Porter 1992). In a given region, one can reasonably predict general migration patterns and the kinds of habitats deer will utilize on a seasonal basis (Dahlberg and Guettinger 1956; Blouch 1984; Tierson et al. 1985). The annual movement of individual or specific groups of deer between habitats is more difficult to specify.

In northern Wisconsin, young forests with abundant browse (e.g., aspen shoots) and open habitats with grassy forage and mast (acorns and other nuts) are preferred by deer during nonwinter seasons. In winter, deer move into forest stands with dense conifer cover to protect themselves from heat loss due to wind exposure and direct radiation to cold night skies. The winter to summer range movements for adult deer averaged 5.6 kilometers in one Wisconsin study, with 90% of the deer moving 12 kilometers or less (Dahlberg and Guettinger 1956). Seasonal migrations elsewhere averaged from 0 to 14 kilometers, depending on the composition and availability of summer and winter range and the severity of winters (Verme 1973; Bratton 1979; Tierson et al. 1985; Murphy et al. 1986). The Forest Service, in the Chequamegon National Forest Plan, conservatively estimated average annual travel distance at 8 kilometers (Sheldon 1984; USDA Forest Service 1986a), though later revised this estimate to "at least 3 miles" (5 kilometers) without documentation (Kick et al. 1991). Although the actual travel distance will vary greatly depending on the individual deer and the area and year in question, 8 kilometers represents a crude but reasonable estimate of the distance over which deer can exert edge effects in our region.

These large herbivores negatively affect populations of certain species of plants, including yew, eastern hemlock, white cedar, and many orchids and liliaceous plants (Miller et al. 1992). Although all of these plants have experienced some level of deer browsing during their evolutionary history, present concern is that current, relatively high populations of deer in many parts of the eastern U.S. may damage populations or reduce their ability to compete with other plants (Alverson et al. 1988; Tilghman 1989; Horsley 1993). High deer populations can also modify habitats in ways significant to vertebrates. Concern that overbrowsing alters forest understory structure in a way detrimental to some bird populations was voiced a decade ago (Kroodsma 1984). Current studies also address the degree to which deer may influence populations of other mammals by competition for food, such as acorns and other mast (McShea and Rappole 1992; deCalesta 1994a,b).

Edge effects due to deer may directly affect the ability of forest habitats to sustain other mammals in yet another important way. The probability of moose successfully recolonizing Wisconsin is quite low at present due to a parasite shared with deer. Forest moose now wander into the northern half of the state from source populations in Minnesota and the Upper Peninsula of Michigan but do not survive here long because they become infected with brainworm, a parasite usually fatal to them but only mildly deleterious to deer. Thus, at present, artificially high deer population levels, creating a high probability that moose will become infected with brainworm, appear to be a major barrier to the recolonization of moose in Wisconsin (Parker 1990).

The Relevance of Edge Effects to Forest Management and Planning

What conclusions can we draw from this brief review of edge effects relative to the protection of viable old-growth stands in eastern forests? In a very real sense, the effects of human activities can be detected in every portion of every continent on earth. In the specific case of temperate forests, few stands are immune to atmospheric or hydrological changes since these factors can operate over such tremendous distances. The range and magnitude of biological edge effects are probably just as great, with changes in the rate and quality of seed, pollen, spore, arthropod, protozoan, bacterial, and viral dispersal probably affecting all remaining forest stands in ways we have only begun to measure.

Negative edge effects caused by vertebrate predators, nest parasites, and herbivores have now been documented in many studies. These animals and birds penetrate forest stands for distances of a few hundred meters up to many kilometers. The magnitude and scale of these edge effects are highly relevant to forest management and intelligent planning efforts. For example, the Chequamegon National Forest staff calculated the acreage of "contiguous forest" for the 9800-hectare project area called Sunken Camp under different edge effect scenarios (Kick et al. 1991). Contiguous forest, in this case, would provide a roughly comparable area of interior forest if allowed to age over the coming decades. Under the planning alternative chosen for the project, 32%

of the area can be considered potential interior forest if no edge effects penetrate more deeply than 200 meters. However, if some edge effects penetrate 500 meters, none of the planning alternatives provides any interior forest habitat because of the distribution of current and future forest timber harvests and wildlife openings.

Using the Forest Sevice's estimate of average annual travel distance of deer in Wisconsin, the center of an old-growth forest stand would have to be at least 8000 meters away from new clearcuts and wildlife openings to remain unaffected by the increased edge effects caused by these forest disturbances. Thus, on average, a circular block of habitat would need a radius of at least 8 kilometers, indicating a minimum area of 20,000 hectares, to include any interior habitat from the point of view of deer herbivory (Alverson et al. 1988). In practice, much larger areas might be necessary to buffer effectively against edge effects due to artificially high populations of deer in our region. One must consider a multitude of factors involved in calculating the extent and severity of effects caused by deer populations generated by any given local timber harvest or wildlife opening. Our point here is merely that wildlife managers who promote deer abundance through the creation of edge habitat are responsible for the effects of these animals on the landscape. In the absence of a detailed and accurate description of the dispersion of their animals over both summer and winter ranges, managers must use existing studies to predict travel distances and the areas affected. In northern Wisconsin and Michigan, these distances are relatively great and should be considered in landscape-level planning efforts for biodiversity.

Summary

In the absence of human intervention, much of the landscape of the eastern United States would revert to forested lands. Through agriculture and silviculture, we counteract this natural tendency of our ecosystems toward large, contiguous blocks of old-growth forest so as to produce commodities such as timber, game, and crops. Most expect that large portions of the landscape will continue to be devoted to management for these disturbed habitats for the foreseeable future. However, the idea that active management for disturbed habitats throughout the landscape yields many benefits but few biological costs represents a dogma now challenged by a growing body of scientific evidence on the negative effect of edge habitats on many native forest species. Furthermore, the great range over which edge effects can be manifested means that managers must think in terms of land units and ecological relationships far larger than those to which they have grown accustomed.

Forest managers and planners, wildlife biologists, and conservation biologists must carefully consider all available data on edge effects in their region and fairly weigh the known or predicted effects on all wildlife for which they are responsible. The many studies of edge effects currently underway and additional disclosure of negative effects produced by traditional management methods will promote a better understanding; but predictive models will, by necessity, weigh heavily in management decisions.

Care must also be taken to avoid unwarranted assumptions or overly simplistic thinking about edge effects. Data on various edge factors should be evaluated with reference to the biology of individual species and only carefully generalized to other species or groups of species, which may differ significantly in their response to the same phenomena.

Negative edge effects have always existed to some degree; but drastic increases in the abundance of brown-headed cowbirds, raccoons, white-tailed deer, and other edge-loving organisms in the present landscape have vastly increased the exposure of remnant forest stands to parasitism, predation, and herbivory. We can, through careful planning efforts at the landscape level, address the changes in two ways: through active management to locally reduce the abundance of these organisms (Diamond 1992; Porter 1992; Alverson and Waller 1993; Garrott et al. 1993) and by providing large blocks of habitat in which the populations of those organisms can be kept near presettlement levels (WFCTF 1986; Solheim et al. 1987; Chapter 11).

6

Hazards of Fragmentation: Area and Isolation Effects

No issues attract as much attention within conservation biology as those connected with geographic scale and distribution: How large should reserves intended to maintain diversity be? How should they be arranged relative to each other? Should they be connected via corridors? Do species common within old growth require areas of some minimum size? Can surrounding younger forests satisfy these area needs?

The last two chapters explored concepts of diversity, habitat patchiness, disruption of historical disturbance regimes, and how edge habitats affect diversity. The magnitudes of these impacts depend critically on the scale at which they occur. If disturbance regimes are only modified locally, or if increases in edges only occur in limited areas, impacts on regional diversity could be minor.

Populations restricted to small areas of suitable habitat often experience hazards that populations distributed more continuously do not. In the first part of this chapter, we consider the particular needs that many species have for habitat areas of some minimum critical size. Other species depend on frequent dispersal among patches of suitable habitat to persist in an area (Chapter 4). Many of these species are vulnerable to isolation in that separate subpopulations do not remain viable when interpatch dispersal is reduced. Considering the separate effects of area and isolation on particular species leads us to then consider how area and isolation interact to affect entire communities.

With a minimal set of assumptions regarding relative rates of immigration and extinction, ecologists have deduced a wide range of consequences with many implications for conservation. Ecologists term this area of research *island biogeography*. Although ecologists' concerns with area and isolation clearly precede and stand separately from the predictions of island biogeography, this body of theory provides a powerful synthesis, including the ability to make quantitative predictions regarding the consequences of habitat loss and fragmentation. In addition, the simple theory provides a starting point to consider more realistic and complex effects. Additional empirical data should further improve our abilities to statistically predict the losses we face as the result of timber harvest activities.

Does island biogeography apply to isolated fragments in forested regions? To answer this question, we critically consider its implicit assumptions and empirical evidence for the accuracy of its predictions. We also consider the extent to which strips of forested habitat may function as biological corridors to unite separate subpopulations.

We conclude that many lines of evidence, including, but not limited to, island biogeography, consistently point toward the significance of cumulative habitat losses and fragmentation as threats to many elements of diversity. While particular mechanisms and rates of species loss remain obscure in many instances, forest managers should now be willing to accept the message that large reserves, and better connected reserves, are necessary to protect elements of diversity sensitive to anthropogenic disturbance. This consensus within the conservation biology community underlies our proposals to establish systems of Diversity Maintenance Areas across public lands (Chapter 11).

Area Needs

When we considered patterns of rarity in Chapter 4 it was noted that most species are rare both in the sense that they occur in only one or a few communities and in the sense that they are scarce within those communities. To understand better how alternative patterns of forest management may threaten gamma diversity, we must consider the particular mechanisms that threaten the persistence of rare populations. These mechanisms revolve around both reduced habitat area and increased isolation, which we consider in turn.

As the amounts of forested area dwindle in some areas, and as old-growth stands continue to be reduced relative to managed forests, many conservation biologists have inquired into the needs various species have for a minimal contiguous area of appropriate habitat. Individual large mammals and birds of prey often have a home range extending over several square kilometers. Large home ranges are especially common among top carnivores that must forage over a wide area in order to find adequate large-bodied prey. Bears, wolves, and cougars are notoriously wide-ranging. Migratory herds of caribou and bison also range widely. For such species, any disruption of, or significant reduction in, suitable habitat spells obvious trouble. An owl that depends on extensive tracts of mature forest to hunt and raise her young may go hungry or fail in fledging those young if her home range is logged over. Even if she survives, her offspring may not be able to find suitable nearby territories if these forests have been cut, possibly forcing them into hazardous attempts to disperse over open lands. More concretely, Beier (1993), in modeling cougar populations in southern California, concluded that contiguous habitat areas smaller than 2200 square kilometers may threaten the persistence of this population. Debates continue over whether the Greater Yellowstone ecosystem at about 65,000 square kilometers is large enough to sustain populations of grizzly bears and timber wolves (Keiter and Boyce 1991).

Herds of migratory ungulates and top predators may not be the only species that need extensive areas of continuous habitat in which to thrive. In southern Wisconsin, for example, red-headed woodpeckers are not sensitive to forest area, while pileated woodpeckers are never found breeding in woods of less than 100 hectares (Ambuel and Temple 1983). Many other plant and animal species occur only rarely, even within their preferred habitat. With such sparsely distributed individuals, these species also may require extensive tracts in order to support an overall population big enough to perpetuate itself. In our region, the round-leaved orchid, for example, occurs only in bogs big enough to provide habitat for an entire population or close enough to other bogs to sustain a viable population. Although it is too late to do definitive studies, passenger pigeons probably also depended on extensive tracts of mast-bearing forest trees to nest successfully, implying that forest losses as well as overhunting may have doomed them (Blockstein and Tordoff 1985).

Wilcove (1988a) has enumerated several other factors that can cause species distributions to depend on the area of habitat patches. These include barriers to dispersal (see below); the absence of particular microsites, or the need for some species like frogs and salamanders to have two or more distinct microsites in proximity to each other; small populations (discussed below); and edge effects (Chapter 5).

Because area needs are so obvious and important, it is necessary to determine what does and what does not constitute habitat for various species (Temple 1988). The answer, of course, depends on the species and the habitat in question. The same areas that serve as large and continuous habitat for some plants and animals will appear subdivided or fragmented from the perspective of others. The openings that constitute barriers to movement and dispersal for salamanders serve as corridors of appropriate habitat for cowbirds and blue jays. As pointed out in the last chapter, edge effects that diminish the value of areas as habitat for some species have no impacts or enhance habitat values for others. This biological relativity makes it important to specify which elements of diversity are being discussed.

As human-disturbed and edge habitats have become more pervasive in the landscape, edge species have become ubiquitous while species dependent on habitats without human disturbance have lost ground. This indicates that biodiversity concerns with area and isolation should focus on this latter class of species, including forest species that depend on old-growth and forest interior habitats. Because we remain ignorant in many cases of what constitutes appropriate habitat and how area and isolation affect particular species, we also face the need to expand and accelerate our research and monitoring efforts (Chapter 7). Only with such information can we ascertain just which areas do, and do not, serve to sustain reproductive success and population viability.

Hazards of Isolation

Population isolation occurs whenever a species' habitat is fragmented or acquires barriers to free dispersal. Isolation may be either physical (e.g., a road) or psychological, reflecting an animal's perception that a given site is unsuitable habitat because of its appearance or the lack of conspecific individuals (as noted for some neotropical migrant birds—Whitcomb et al. 1981). In many cases, the deleterious effects of isolation are as important as the effects of decreased area. Furthermore, these isolation effects are often difficult or impossible to separate from the effects of decreased area. Nevertheless, we discuss them separately here because they constitute a distinct class of threats.

Populations of most species are geographically structured, consisting of multiple subpopulations that are relatively independent from one another (Chapter 4). Subpopulations with an appreciable reproductive surplus often act as source populations, sending out a stream of emigrants that may succeed in entering existing populations or founding new subpopulations. Other subpopulations on the landscape represent net population sinks in that by themselves they would tend to decline. These sink populations depend on the regular influx of individuals from other populations in order to persist. While it is tempting to think of them as ecologically marginal, they probably play important ecological roles by providing opportunities to colonize more distant sites and by acting as stepping stones to knit metapopulations together. It also seems likely that whether populations act as sources or sinks could shift with changes in conditions. If such shifts occurred often, the persistence of the metapopulation might depend on retaining an adequate network of occupied patches. Because metapopulation dynamics occur commonly, demographic exchange and subpopulation proximity may often foster population persistence (Gilpin and Hanski 1991).

If populations of many species depend on the sporadic, but continuous, input of immigrants from other populations, isolation represents a significant threat to diversity. Populations that are not self-sustaining need to be periodically rescued by the arrival of new plant or animal colonists from elsewhere. The overall growth rate of the metapopulation and its ability to persist may hinge critically on these exchanges. In such cases, habitat fragmentation interferes with natural population dynamics by isolating populations from one another, preventing demographic and genetic exchange. Isolated populations are often further compromised by increased edge effects, turning source populations into sinks.

Isolated populations are also more likely to go extinct because they are smaller. Small populations are inherently more likely to experience biased sex ratios, uneven age distributions, and the accidents of high and low survival and reproduction, termed *demographic stochasticity*. The importance of these chance events increases in small populations for the same reason that results from a few coin tosses vary more than the sum of many coin tosses (what mathematicians term the Law of Large Numbers). Thus, a population might experience several consecutive years of high mortality or low reproduction, driving it as low as 10 or 20 individuals, at which point almost any other biotic or abiotic stress or inbreeding might extinguish it. In continuous populations, individual subpopulations are often rescued by immigrants from adjacent subpopulations, buffering local fluctuations and allowing persistence (Chapter 4). As subpopulations become more isolated, however, immigrants only arrive rarely. If subpopulations disappear more often than they are recolonized, the metapopulation starts to decline and an increasing proportion of patches of suitable habitat will remain unoccupied. Thus, the spatial structure of populations and their proximity to travel corridors and barriers greatly affect their overall performance and persistence.

If new immigrants do not arrive regularly, populations also become genetically isolated. The genetic isolation of small populations usually brings the random fixation of particular genetic combinations, termed *genetic drift* (Crow and Kimura 1970; Futuyma 1986). While drift within subpopulations sometimes may play a critical role in long-term evolution (Wright 1931, 1982), it does so only when populations are able to at least occasionally exchange immigrants and when opportunities to establish new populations exist in the landscape. Without these opportunities, drift represents a loss of genetic variation that may increase the vulnerability of subpopulations to novel biotic threats or to other changes in their environment.

Genetic isolation also increases the likelihood of mating between relatives, termed *inbreeding*. While effects of inbreeding vary (Templeton and Read 1984), populations that have historically outcrossed often experience pronounced inbreeding depression when they start to inbreed. The effects of inbreeding depression are manifest in many traits, including misshapen sperm, an increase in birth defects, developmental abnormalities, increased sensitivity to disease and parasites, or reduced growth and reproductive ability. In mammals, these effects can greatly reduce fitness when matings occur between cousins or half-sibs (Ralls et al. 1979; Ralls and Ballou 1982). Although plant species often persist as small populations and tolerate close inbreeding, even habitually inbred plant populations often experience inbreeding depression (Waller 1993a). In a significant study involving royal catchfly in remnant prairies, Menges (1991) found that germination plummeted from 85% in populations above 150 to less than 25% in very small populations. Thus, genetic effects also remain a major focus within plant conservation (Falk and Holsinger 1991).

Matrix Effects

In addition to the effects of reduced demographic persistence and increased inbreeding and genetic drift, fragmented populations also often face alterations in the surrounding matrix that threaten their persistence. Lands surrounding most eastern forest stands are roaded and include a high proportion of early successional and edge habitats. This may compromise the ability of these forested stands to sustain diversity by exposing them to a rain of propagules from weedy species as well as high densities of predators, parasites, and herbivores (Chapter 5). Such changes in composition (together with direct human management activities such as fire suppression) greatly alter the disturbance regimes that prevail in the landscape with similarly profound effects on remaining patches of habitat (Chapter 4).

Within the continuous matrix of old growth that made up the presettlement forests, forest interior plant and animal species encountered few insurmountable barriers to dispersal and readily recolonized disturbed areas as they succeeded into old growth. In areas where older stands are small and widely separated from one another, forest-interior or disturbance-sensitive species, including salamanders, certain forest herbs, and some insects, experience difficulty in migrating or dispersing among appropriate habitat patches. In addition, the proliferation of roads has tended to favor weedy species that disperse well along roads and in other human-disturbed sites.

The presence of open roads in an area may transform it from a suitable to an unsuitable habitat for many species (Bennett 1991). Access by humans to most forested regions occurs via roads, making road density a useful predictor of an area's ability to retain certain species such as wolves and bears sensitive to the presence of humans (Harris and Atkins 1991; Wilcove 1989; Noss 1991a). Wolves in northern Wisconsin and Minnesota rarely occur in areas with more than 0.9 miles of open roads per square mile, and populations are denser where road densities are lower (Thiel 1985; Mech et al. 1988). Brocke et al. (1988) also note that cougar avoid active roads and that moose kills are higher in areas accessible because of logging roads.

Black bears now occupy only 5–10% of their original range in the southeastern U.S., mostly in relatively undeveloped areas (Pelton 1985). On other lands, due to logging operations and access roads, they simultaneously encounter fewer nut-bearing mast trees and more humans (including poachers eager to sell their internal organs to buyers in the Far East). Radio-collared black bears die most often by being shot by hunters (Jonkel and Cowan 1971; Rogers 1987), some of whom may use radios in hunting them. In the Adirondacks, the number of bears shot legally was closely correlated with road density among 12 mostly forested counties, where a 10-fold increase in road density brought a 10-fold increase in kills (Brocke et al. 1988). Human population density and forest cover were less predictive of the kill rate.

In addition to providing access for humans, roads act as barriers for some species and avenues for invasion for others. Weedy plants such as purple loosestrife often spread via the ditches along roads or attached to vehicles. Roads also increase the amount of edge habitat present. Some rare native species benefit from open patches along roads, including grass-of-Parnassus in Door Co., Wisconsin, flat oatgrass in Grand Isle, Michigan, and New England violet in Voyageurs National Park, Minnesota. Nevertheless, the number and significance of deleterious effects associated with roads clearly indicate that road densities should be reduced in and around some areas intended to conserve biodiversity.

Given that so many plant and animal species are rare and that many of these are sensitive to area or isolation effects, it behooves us to attempt to devise general principles that can be used to predict the effects of lost habitat area or increased habitat fragmentation. Efforts to protect such species must also address the entire ensemble of the threats they face, rather than only one or two. Is it possible to predict at least in a statistical sense how many species are likely to be threatened by any given reduction in area? It would be even more useful to be able to predict exactly which species might become threatened—might this be possible as well? While it is usually difficult to predict specifics, ecologists are now confident about predicting general relationships between area and diversity and the aggregate losses likely to follow habitat loss. To pursue these questions, we turn our attention to the seemingly arcane, but actually fundamental, area of ecological science known as island biogeography.

Island Biogeography

About 40 years ago, biogeographers began to investigate patterns of diversity on oceanic islands. By compiling lists of species from islands of different sizes and at varying distances from other islands or the mainland, they began to elucidate the general factors that determine the numbers and kinds of species found on islands (Preston 1962a,b; MacArthur and Wilson 1963, 1967). From these studies emerged MacArthur and Wilson's theory of island biogeography, which views species diversity on islands to be the result of a dynamic equilibrium between immigration and extinction. Diamond (1976) and Wilcox (1980) then pioneered the application of this very general body of theory to the field of conservation biology.

As larger and larger areas are sampled, one encounters more and more species, but the rate of finding new species declines. When one plots logarithms of both species number (S) and area (A), one usually gets a line, reflecting a power relationship:

$$S = c A^z,$$

where z is the slope of the line on the log–log graph and c is a scaling constant. As one might expect, the slopes of these lines differ between sampling on isolated islands ($z = 0.12 - 0.17$) and sampling on the mainland or bigger islands ($z = 0.2 - 0.35$; Fig. 6-1). If z is 0.3, a 10-fold increase in area doubles the number of species present, a generalization first noticed by Darlington (1957). Isolated islands have fewer species at any given size and also show sharper declines in species number with decreasing area than mainlands. Assuming a particular pattern of relative abundance for

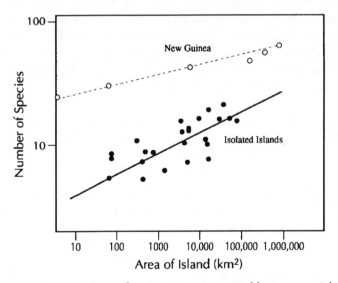

Fig. 6-1. Species–area relations for pomerine ants compared between a mainland site (New Guinea) and more isolated Moluccan and Melanesian islands in the region. [From A. N. Wilson (1961), *The American Naturalist* 95: 169–193; reprinted by permission of the University of Chicago Press, © 1961.]

species (e.g., the common log-normal distribution) allows one to then derive the species–area relationship (Preston 1962a,b; MacArthur and Wilson 1967).

Because most habitats are patchy to some degree, the concept of insularity is remarkably general. Almost any patch of habitat isolated within a different type of habitat can be considered an ecological island: lakes or bogs within forests; fallen logs in a forest; wood lots surrounded by farmland; or patches of open grassland within a forest. Mountaintops often act as islands for the alpine species that occupy them but not surrounding lowlands. Of course, what is the "island" and what is the "ocean" depends on the habitat requirements of the particular creature in question. In general, our concern here is with stands of mature forest embedded in a "sea" of younger, mostly early successional tree species.

The rate at which new species immigrate to an island or patch can generally be expected to decrease with increasing distance from the mainland or other similar patches of habitat. This rate will also tend to decrease with increasing numbers of species already present on the island. This occurs because more new immigrants will belong to a species already there and because similar species already present may compete with immigrants of new species. Rates of extinction also vary predictably. On small islands, we expect species to go extinct faster because smaller areas support smaller populations and smaller populations are more likely to go extinct. Similarly, as species number increases, extinctions become more likely, both because there are more species to go extinct and because more competitive conditions are likely to prevail. (Actually, extinction rates might actually decrease at first with increasing number of species in cases of primary succession, as the presence of some species may facilitate establishment by others—see MacArthur and Wilson 1967, p. 50.)

From these simple ideas follow several straightforward conclusions. Most immediately, we see that islands should support some equilibrium number of species where the rate at which new species colonizing the island (or habitat patch) are balanced by the rate at which existing species go extinct (Fig. 6-2). At this dynamic equilibrium, the rates of immigration and extinction cancel, yet both continue—a process termed *turnover*. Larger islands support more species, mainly because extinction rates are lower. Islands closer to sources of immigrants also support more species, mainly because the rate at which new species colonize the island is higher. Rates of turnover should be higher on smaller islands due to their higher extinction rates, and on near islands due to their higher rates of colonization. The theory has been extended to make more quantitative predictions, but these simple conclusions represent its essence.

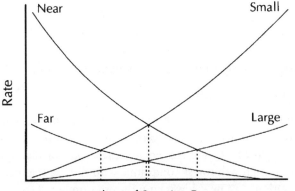

Fig. 6-2. Graphical depiction of how declines in the colonization rate and increases in the extinction rate with increasing numbers of species on islands lead to a predicted equilibrium number of species on islands. Note that increasing distance lowers the immigration rate and increasing island size decreases the extinction rate. [After MacArthur and Wilson 1963.]

As with any theory, the limitations and simplifications of island biogeography should be kept in mind. The theory makes statistical predictions regarding the (average) total number of species and their (average) turnover rates. In this sense, it resembles many other theories, such as quantum mechanics. In particular, note that no special assumptions have been made regarding which species are more likely to colonize islands, the area needs of particular species, or the sensitivity of different species to extinction via isolation. Island biogeography is a general theory that assumes little regarding the details of immigration and extinction and predicts in a statistical sense the number of species present at equilibrium and the rate at which colonization and extinction balance. In its simplest form, however, it does not predict which species will colonize, which will go extinct, or which will be retained at equilibrium. Nevertheless, MacArthur and Wilson (1967) realized that these events would be nonrandom and developed the theory of r- and K-selection to explain how natural selection tends to favor different suites of life history traits in accord with the differential opportunities various species have for colonization of new sites relative to persistence in established sites.

Island biogeography makes an important distinction between land-bridge islands that were once connected to the mainland and islands that arose *de novo* in isolation from sources of colonization (Diamond 1972, 1973; Terborgh 1974; Wilcox 1978, 1980). Land-bridge islands have substantially more species than would be predicted by their size, as they represent former "continental" areas that have become isolated and are gradually losing species. Such losses are expected to continue until a new equilibrium is reached in accord with their reduced area and degree of isolation. The period of time it takes for the island to lose its "excess" species is termed its *relaxation time* by analogy with physical oscillators. Fragmented natural communities derived from larger original blocks of habitat are thought to resemble land-bridge islands in that their smaller size and increased isolation precipitate losses of diversity over time.

The concept of relaxation time is particularly relevant to small, isolated fragments of old growth. Predictably, certain species that now occur in these fragments will be lost if these areas remain small and isolated. Species with high fidelity to old-growth forest conditions within these fragments are undergoing "relaxation" to a new, lower equilibrium number of species in a pattern similar to that observed on land-bridge islands. In these cases, a population's persistence may be enhanced if its dynamics are relatively slow (e.g., long-lived plants). Slowed losses are important for two reasons. First, they mask the immediacy of the threats and lull many into thinking that because these species have not yet gone extinct, they are safe. Second, they aid conservation efforts if opportunities are provided for these stands to serve as inocula to colonize the surrounding matrix of maturing stands.

Many studies of oceanic and habitat islands strongly corroborate the basic theory and suggest that it could have considerable utility in conservation planning (e.g., Reibesell 1982; Wilcox et al. 1986; Shafer 1990). For bird and lizard populations on land-bridge islands, small islands have indeed lost relatively more species than larger islands, displaying the faster turnover rates predicted by the theory (Diamond 1972; Terborgh 1974; Wilcox 1978). Direct estimates of extinction rates have also been made for insects and lizards on conventional islands (Rey 1981; Schoener and Schoener 1983). Although often displaying slower dynamics, the distribution of plants on islands in Lake Michigan and off the coast of Maine also fit predictions of island biogeography (Nilsson and Nilsson 1982; Forzley et al. 1993; Judziewicz and Koch 1993; T. Gibson unpubl. data). Small islands consistently show lower diversity and higher extinction rates. These results all support the idea that smaller fragments of remnant habitats face the greatest risks of losing species in the near term.

How Applicable Is Island Biogeography to Conserving Forested Habitats?

Given its generality and the multiple tests that have borne out its basic predictions, the theory of island biogeography appears rather widely applicable to situations in which specialized plants and animals adapted to certain types of habitat occupy areas surrounded by inappropriate habitat. Furthermore, it provides explicit and testable predictions regarding the number of species that can persist on islands of various sizes and degrees of remoteness. In fact, island biogeography has served as a catalyst within ecology to focus attention on the dynamic processes that underlie species coexistence, distribution, area effects, and community composition. Thus, it is only natural that conservation biologists have eagerly sought to apply ideas from island biogeography to a variety of conservation issues.

In addition to Diamond's (1976) seminal article, several authors have deduced that area and proximity represent fundamental properties important for predicting the ability of natural areas to sustain biodiversity (Forman 1974; Wilcox 1980; Burgess and Sharpe 1981; White et al. 1983). In fact, two books specifically examine how island biogeography could be used to design nature reserves (Shafer 1990) and conserve sensitive species in northwestern forests (Harris 1984). Biologists concerned with the implications of island biogeography for conservation frequently emphasize the need to protect large areas of land and to link or distribute these in such a way as to foster immigration among them (Diamond 1976; Wilcox 1980; Soulé and Simberloff 1986). Interestingly, these same recommendations could be (and have been) made purely on the basis of known edge, area, and isolation effects, without recourse to island biogeography. Nevertheless, the theory has helped spawn further interest in habitat fragmentation and edge effects, even when these do not depend on the area effects central to island biogeography.

Newmark (1987) provided a dramatic application of island biogeography in his analysis of species losses in the western U.S. national parks. In considering patterns of 42 extinctions of larger mammals among 14 parks of varying size and age, he found that larger parks experienced far fewer losses than smaller parks, and that area accounted for 56% of the variance in how many species were lost. When park area was controlled, park age also appeared to influence the rate of loss. The analysis excluded extinctions directly caused by humans but included indirect human-related impacts. Thus, two independent predictions from island biogeography appear to be met in this study. While many biologists were not surprised to read yet another corroboration of the theory, many were shocked that losses of this magnitude have occurred within a century in some of our largest national parks specifically set aside to conserve natural values and wildlife.

Despite these successes, some criticize the application of island biogeography to temperate forests, often basing their criticisms on the fact that particular assumptions of the theory may not be met in terrestrial habitats. The analogy between terrestrial habitats and oceanic islands is imperfect in that it depends on the kinds and degree of interactions between a habitat island and the matrix of different habitats surrounding it. Even oceanic islands like

Hawaii interact with the ocean via strong meteorological and biotic interactions. For example, the ancestors of most of the plants found on such oceanic islands originally colonized by floating long distances from other island or continental sources (Carlquist 1974). In terrestrial ecosystems, habitat patches of different type often interact to a high degree with the surrounding biotic matrix. This complicates the simpler aspects of the theory in that ecological relationships influence rates of immigration and extinction. McCoy (1982) lists the criteria for a formal proof of the theory in its application to terrestrial systems.

One assumption that many have questioned concerns the degree to which surrounding younger forest stands ameliorate the isolation of older stands by reducing edge effects and effectively buffering or enlarging each old-growth fragment. Such stands do provide forested conditions that appear to provide at least some of the characteristics of older stands, including shade. The extent to which terrestrial islands of old growth mimic true islands clearly depends on which species one considers. Younger forest matrices benefit some species (e.g., deer) but appear to be of little or no benefit to many forest interior and area-sensitive species. In addition, younger forests differ from mature forests in species composition, soil and stand structure, prevailing disturbance regimes, and many physical conditions. As Curtis's (1959) results on Wisconsin understory plants showed, most species are quite restricted in the range of habitat conditions they will tolerate. Younger forest matrices may even accentuate threats to elements of diversity in some circumstances. Seedlings of Canada yew, for example, face bleak odds of surviving when surrounded by young forest supporting moderate to high deer densities. In such cases, far from reducing losses, a younger forest matrix may directly threaten elements of diversity surviving on remnant patches. Within the island metaphor, deer are analogous to oceanic sharks able to climb onto shore to ravage inland dwellers. Given these external threats (Janzen 1983, 1986), it is both naive and misleading to claim that younger forests will categorically ameliorate threats to mature forest species.

Discussion has also raged at times over when and whether a single large reserve might protect diversity better than several small reserves, an issue often referred to by its acronym: the SLOSS (single large or several small) debate (Simberloff and Abele 1982). Small reserves can be dispersed to cover existing remnant patches of old-growth and other biotically unique and important habitats, making them an important component of any conservation scheme. Such a collection of small reserves will also often contain a higher number of species, at least initially, than a single larger reserve. No reserve is fully isolated, however. What is important is how well the reserves function in conjunction with one another and with surrounding landscape elements over the long term to best maintain regional (gamma) diversity. In this regard, large reserves appear much better able to independently sustain viable populations of area-, edge-, and isolation-sensitive species. They also are less sensitive to external influences and therefore simpler and cheaper to manage (Alverson and Waller 1993). Any network of smaller reserves will therefore have to be carefully designed not only to capture but also to sustain regionally viable populations of sensitive species. Soulé and Simberloff (1986) agreed on these general issues regarding the ecological role of large reserves. These arguments support our contention in Chapter 11 that at least some large old-growth reserves are needed to adequately protect forest biodiversity.

Others question the assumption that immigration rates vary enough in terrestrial habitats to produce the downward sloping immigration curve assumed in island biogeography (Brenneman and Eubanks 1988; Hunter 1990). These criticisms extend particularly to vagile groups like migrant birds. While this objection may have some validity, there is evidence, even for many bird species, that immigration rates do vary significantly among habitats of varying size. Pileated woodpeckers and many neotropical migrants discriminate in favor of larger habitat areas (see above). It also appears evident that overall immigration rates have declined significantly for many forest interior and old-growth species with forest cutting and fragmentation. While these are important issues that deserve further study, they do not challenge the essential insight of island biogeography that numbers of species depend on the dynamic processes of immigration and extinction.

In some instances, criticisms of island biogeography appear to reflect a lack of appreciation for the nature of theoretical constructs, all of which represent simplified caricatures of biological reality. Thus, although the theory appears only partially able to account for the distributions of species among forest fragments (Wilcove et al. 1986), no conservation biologist we know of has ever argued otherwise. Furthermore, these debates have actually served to strengthen island biogeography and extensions thereto by focusing increased research efforts on obtaining empirical data on rates of immigration, extinction, and turnover.

How Do Immigration and Extinction Rates Vary?

MacArthur and Wilson (1967) and others interested in the predictions of island biogeography (e.g., Lovejoy and Oren 1981) have long been aware that species vary in their propensity to colonize new areas and in their ability to persist in habitats of different sizes. Some species are vagile and itinerant, while others are quite sedentary. Similarly, some populations appear precarious, with regular extirpations often due to small population sizes (Elliott

1986). Is it possible to extend island biogeography to include differences in immigration ability and persistence rather than assuming all species to be similar and interchangeable? If so, the theory might be capable of predicting not just the number but also the kinds of species that can be expected to occupy differently sized patches of habitat.

Obvious differences in size, mobility, and behavior surely influence the likelihood of colonizing dispersed sites. In general, ruderal or weedy species are best adapted to continually colonize new areas because they are able to take advantage of various kinds of disturbance (Chapter 4). In contrast, species typical of late successional or relatively undisturbed habitats are commonly more resident and often only have the physical ability to disperse locally. They may also be psychologically averse to crossing openings.

Diamond (1975) significantly advanced biogeographic theory by developing the idea of incidence functions. He documented the presence of differences in colonization ability among birds and the role these differences play in determining the distributions of birds among New Guinea and the surrounding Pacific islands. In particular, he found that some species were especially adept at colonizing even small and remote islands while others were only found on islands above a certain size. He termed the vagile and correspondingly widely dispersed species "supertramps" because of their ability to wander and survive. Many of these species were absent from larger and closer islands, presumably reflecting their inability to compete with more resident birds once those became established. The existence of such differences among species in colonization rates and coexistence implies that these bird communities have predictable "assembly rules" that depend on island size. Bird species in temperate North American forests also display characteristic incidence functions in that their occurrence often depends on habitat area (Fig. 6-3).

If extinction also is deterministic, we would expect habitat patches of decreasing area to lose species in a predictable order. The result of such "disassembly rules" would be that successively smaller habitat patches would contain successively nested subsets of the species present in larger patches. Bird communities in east central Illinois and bird and mammal communities on mountains in the Great Basin region show statistically significant nested patterns of species distributions (Blake 1991; Cutler 1991). These studies provide strong evidence that some species are far more extinction-prone and area-sensitive than others, an issue of long-standing concern to conservation biologists (Terborgh 1974). The patterns are not absolute, however, and often contain "holes," where a habitat is predicted to contain a species that it does not, or "outliers," where a species normally present in larger patches occurs in a patch smaller than expected. Interestingly, Cutler found that the

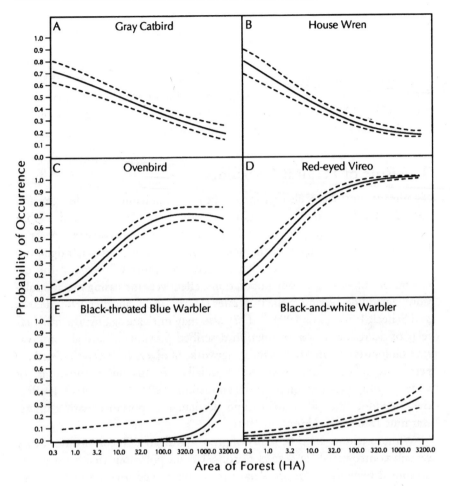

Fig. 6-3. Probabilities of detecting various bird species as a function of contiguous forested area, as sampled at random points in forests of various sizes in the middle Atlantic states (from Robbins et al. 1989). Typical resident birds or short-distance migrants (A and B) generally decline in frequency with forest area, whereas many forest interior neotropical migrants (C–F) become more common in forests of increasing size. These probabilities resemble the "incidence functions" recognized by Diamond (1975) for birds that differ in their ability to colonize and persist in islands of various size.

nested subsets of mammal distributions tend to have holes while bird distributions tend to have outliers. Such deviations might be expected given that many mammals are relatively sedentary but vulnerable to local accidents, while many bird species temporarily occupy sites on flights between more appropriate habitats. Specialized old-growth species of low vagility that are slow to recolonize will also display holes in their expected distributions in cut-over landscapes, e.g., some herbs and ground beetles (Duffy and Meir 1992; Niemelä et al. 1993).

These regularities in immigration and habitat occupancy greatly enhance our ability to predict which species depend the most on area or are otherwise vulnerable to extinction. Again we see the value of ecological theory in extending the ability of conservation scientists and forest managers to anticipate the consequences of changes in land management.

The Role of Habitat Corridors

Significant controversy also exists over the roles that habitat corridors can and should play in maintaining regional biodiversity (Saunders and Hobbs 1991). Some biologists have sought to identify those particular aspects of landscape structure that allow habitat strips to act as corridors (Bridgewater 1987). Proponents point out that effective corridors will expand the exchange of individuals among populations, effectively increasing metapopulation size and decreasing extinction rates (Harris and Gallagher 1989; Harris and Scheck 1991; Noss 1991c). They also may mitigate against the other effects of decreased area and isolation described previously. Several conservation biologists are already designing networks of diversity reserves knitted together with corridors intended especially to sustain populations of wide-ranging top carnivores (Harris and Atkins 1991; Noss 1991c). If effective, corridors are likely to assume particular importance with climatic warming (Hobbs and Hopkins 1991).

While corridors are potentially of great significance to conservation, they remain controversial for several reasons. Critics point out that most of the presumed benefits of corridors have yet to be proved, that corridors might also promote the spread of diseases or edge-loving species, and that corridors that bisect developed landscapes can be very expensive (Hobbs 1992; Simberloff et al. 1992). Given their many potential benefits and these several remaining uncertainties, the use of corridors will doubtless remain controversial until further research clarifies their utility (Nicholls and Margules 1991). In the meantime, biologists will have to use their judgment regarding the region's species and communities to decide when and where corridors are justified. In our opinion, corridors are likely to turn out to be critically important for sustaining species in many regions.

Summary

The fragmentation and harvesting of temperate forests have greatly diminished the area of remaining old-growth and forest interior habitats, especially in eastern North America. As these habitats have been reduced in size and increasingly isolated, elements of diversity dependent on the particular conditions and disturbance regimes they support have become scarcer. Many species have already been lost. Other losses continue, as with the massive declines in neotropical migrant birds witnessed in recent decades. Still other losses are not yet evident because species with long life spans only slowly respond to habitat loss and because relaxation to lower numbers of species in newly isolated habitats takes time.

Of course, species rarely face one of these problems at a time. Forest fragmentation changes disturbance regimes at the same time it increases edge habitats, reduces the area of core stands, and increases isolation among stands of mature forest. Because these factors frequently act in combination, their total impacts are likely to be more severe than the sum of their individual impacts. It will also be quite difficult to determine in many particular cases which of these factors is most significant. Contemporary discussions of threats to diversity in temperate forests stress the distinct and cumulative threats that exist for various elements of diversity (Wilcove et al. 1986; Wilcove 1988a,b). Understanding the mechanisms contributing to these losses will help us devise better strategies to protect remaining species and restore those that have declined or suffered local extirpation.

While mechanisms of endangerment will long continue to be the focus of research and discussion among conservation biologists, forest managers should concern themselves primarily with two consistent and congruent messages that have already emerged from these studies:

- Habitat loss and fragmentation are significant threats to many elements of diversity, and

- Big reserves (or well-integrated and well-connected networks of smaller reserves) will be necessary to sustain sensitive elements of diversity over the long run.

Managers should not interpret uncertainties about which particular factors threaten which particular elements of diversity as an indication of uncertainty over these fundamental lessons of conservation biology or as excuses to postpone implementing conservation programs based on them.

7

Tracking Diversity:
Biological Inventory,
Research, and Monitoring

Conservation biologists spend much time resuscitating species and communities on the brink of extirpation and attempting to understand mechanisms of population endangerment. Indeed, the field is preoccupied with understanding the pathologies of low or declining populations and treatments to restore population and community viability. We pay less attention to diagnosis. When do we recognize that a population, species, or community type is in trouble? What vital signs should we monitor to assess ecosystem health? How can we reliably assess how management is affecting sensitive components of diversity? Answers to these questions are of obvious importance to land managers charged with conserving biotic resources.

Forest managers depend on biological inventory and monitoring to assess how well diversity is being maintained and how management practices affect particular elements of diversity. Given the diversity of the biotic resources they manage and the ecological complexity of forests, these are daunting tasks. No one would expect the manager of a nuclear power plant or the pilot of a jumbo jet to succeed in his or her job without considerable training, expert technical assistance, and sensitive and reliable instruments to indicate the status of key subsystems. Today's teams of forest managers, however, are expected to succeed at even more complex tasks with rudimentary knowledge, limited technical assistance, and imprecise and untested monitoring tools.

Assuming they are wisely chosen and closely monitored, ecological indicators are the "canaries in the mine shaft" that signal potential trouble for sensitive elements of diversity. They can provide early and accurate feedback regarding how management is affecting populations, species, and communities across the landscape. It is therefore critical that the methods used to track indicators be consistent and reliable. Monitoring should also be as economical as possible. Because it is impossible to track all, or even a substantial fraction of, the biological elements in a forest, managers routinely rely on monitoring a small subset of the species and communities they oversee. Potential ecological indicators range from the genetic composition of individual populations through indices of community and ecosystem performance. Most managers, however, rely on particular indicator species.

The use of indicator species to guide forest management was codified under the 1976 National Forest Management Act and associated federal regulations that require each National Forest to choose and monitor an appropriate set of indicator species. By monitoring these species, forest managers assume they can detect incipient changes in habitat conditions and forestall species losses through appropriate changes in management. The Forest Service faces unique opportunities to integrate research and monitoring with management as it supports a highly qualified scientific staff in its research branch and has historically embraced a tradition of scientific forestry. Despite such advantages, however, the Forest Service has lagged in implementing effective monitoring schemes and applying its research to management. In fact, monitoring schemes initially devised for the Wisconsin National Forests were based on a limited and often inappropriate set of biological indicators that limited the utility of the data and had the effect of favoring more intense methods of management (see Chapter 13).

In this chapter we explore the biological principles and methodological tools used to inventory and monitor biological resources. After reviewing historical uses of the indicator concept, we explore the need for biological survey data and the assumptions and trade-offs implicit in choosing indicators intended to be simultaneously effective, reliable, and efficient.

With this background, we then present a list of principles that we feel should guide research and monitoring efforts. We emphasize that inventory and monitoring should occur across many scales and that different indicators provide different, but often complementary, information that should be properly synthesized. We also stress the need to integrate research and monitoring more closely with management. Finally, we provide a list of specific research questions to demonstrate the relevance of research to managers intent on protecting diversity. The number of basic questions that remain unresolved supports our earlier contention (Chapter 2) that we still have much to learn regarding how elements of diversity are affected by forest management. Our recommendations regarding monitoring should be considered provisional as these methods continue to evolve with our understanding of how environmental changes affect elements of diversity.

Uses of Biological Indicators

Biotic indicators are generally used where the presence or state of an organism may be simpler or cheaper to monitor than the physical or chemical process of interest (National Academy of Sciences 1979). Over 100 years ago, European botanists noted that certain lichen species had disappeared from industrial cities (Ferry et al. 1973); lichens are still considered reliable indicators of air pollution (Nash and Wirth 1988; USDA Forest Service 1993). Similarly, invertebrates are employed to indicate water quality (Phillips 1980; Hilsenhoff 1982), and mammals have been used to monitor metal contamination (Wren 1986). More recently, scientists have begun to use satellite imagery to monitor crops and forests across entire regions, including losses due to logging and fires.

The presence or abundance of particular species has long been used to infer the presence of particular community types or the condition of agricultural land or pasture range (Clements 1916; Stoddart et al. 1975). In fact, the concept of indicator species originated with European plant ecologists who sought out particular species to characterize community types. Their goal was to choose species faithful to particular community types. Indicator species in this sense often reflect soil and site conditions, making them suitable indicators of site quality for silviculture as well. Kotar et al. (1988), for example, developed a sensitive system for assessing forest site conditions in the upper Midwest based primarily on herbaceous indicators.

More recently, indicator species have been used to reflect the overall ecological status of the habitats these species occupy (Salwasser 1987; Noss 1990; Goldsmith 1991; Spellerberb 1992). Conservation scientists use the population density or reproductive success of particular species to indicate whether ecological conditions are favorable for that species. A population's decline or demographic failure is assumed to signal deterioration in habitat quality or abundance. Vertebrates have the virtue of being sensitive to particular environmental toxins. They concentrate fat-soluble toxins such as chlorinated hydrocarbon pesticides and PCBs, making carnivores at the top of the food chain quite vulnerable to poisoning (e.g., the losses of peregrine falcons due to eggshell thinning).

The assumptions involved in using indicators are also worth scrutinizing. Typically, populations of only a few vertebrate species are monitored, leading some to call for monitoring a taxonomically wider set of indicator species (e.g., Cairns 1986). Noss (1990) extended this idea by arguing that ecological indicators should include not just populations of particular species, but a range of ecological structures and processes. He promotes a top-down approach that begins with coarse-scale inventories of the distribution of plant and animal communities and species across the landscape, then focuses intensive monitoring on those communities and elements of diversity judged to be most at risk. Such a hierarchical approach takes advantage of our current understanding of how species composition and ecological processes interact. We endorse his approach and stress that its value rests squarely on accurate and sufficient inventory data.

The Need for Systematic Inventories

Informed management of complex forest ecosystems logically begins with a thorough knowledge of their biological composition. This might take several forms. To most people, a biological inventory implies lists of species present in an area, accompanied perhaps by a survey of their approximate abundance or distribution. Such surveys are not trivial, even for mammals and birds, but nevertheless provide significant useful information. These data assume particular importance when they are accessible via a comprehensive regional database containing information from museum collections and herbaria. The Nature Conservancy has initiated such databases for the rare and threatened flora and fauna of most states via its Natural Heritage program. The systematics community is participating actively by developing additional on-line databases to include comprehensive information on plant and animal collections.

Gap Analysis

Once assembled, geographic data should be systematically examined to assess which community types are least well represented in the region by being protected within existing natural areas such as Wilderness, Parks, Research Natural Areas, etc. Such analyses should be integrated across entire regions and all ownerships and attempt to assess the extent to which surrounding private lands act to enhance or reduce the persistence of threatened elements of diversity present within public lands. Those community types deemed to be underrepresented or threatened by fragmentation and isolation should then become a priority for reserved status and more intensive monitoring.

In the U.S., Scott et al. (1991) have pioneered using the geographical ranges of birds and mammals present in a region to make judgments regarding priorities for extending natural reserve networks. Their *gap analysis* approach has now been applied in several western states. Extending the technique to include other groups and adding more weight to rare species and known area dependencies would enhance its effectiveness further (Waller 1991). It might also prove useful to emphasize the geographic ranges of regionally rare or endemic species rather than the ranges of all species present in a region.

These approaches require a great volume of high quality data, emphasizing the need for accurate and complete inventories and continued monitoring. Because complete and reliable information on the ranges of individual species will likely remain scarce, it is usually simpler to consider the distribution of vegetative communities present in an area. Every forest manager should have detailed and accurate maps of the forest community types that exist within his or her jurisdiction and an appreciation of their nature, distribution, and special needs. Biologists, who usually base their recognition of community types on the tree species that dominate the canopy, may also include information on soils, herbaceous species, and approximate seral stage. Unfortunately, community classification schemes are inevitably arbitrary and vary in specificity, leading to considerable heterogeneity among regions and among classification schemes. Thus, several groups in the Midwest have called for the establishment of some uniform system of community typing (Albert 1993; Crow et al. 1993). Foresters have also embraced the related concept of "habitat type," where soil conditions or a set of indicator plants are used to predict the potential forest community or the site's potential for growing timber (its "site index"—Kotar et al. 1988).

Geographical Approaches

To anticipate and predict the indirect and cumulative effects of management, we also need to extend and refine our use of geographical planning and analysis tools. The diffuse, but potent, effects of forest fragmentation are now known to threaten many edge-, area-, and isolation-sensitive species (Chapter 6). These elements of diversity depend on geographical aspects of the landscapes they occupy, making it important to develop and apply geographical planning tools. Here, forest managers will be aided by ongoing developments in geographical information systems (GIS) and landscape ecology (Turner 1987; Turner and Dale 1991). It is now possible, for example, to combine data from field surveys and community maps with knowledge of how particular biotic elements respond to landscape elements to predict regional persistence in response to various land management decisions.

Thus, as our knowledge has increased, we have become more aware of the need for accurate inventory data coupled with an understanding of geographic interactions. Many distinct lines of evidence support "coarse-filter" approaches to conserving diversity by protecting ecological processes over relatively large areas (Chapter 10). In the U.S., the Wildlands Project has concluded that it may be necessary to protect an ambitious continental network of large reserves, buffer areas, and corridors in order to fully protect diversity (Foreman 1992; Noss 1992).

Assumptions and Conundrums of Monitoring

Choosing any subset of ecological indicators to reflect conditions for entire communities involves several assumptions that are best made explicit. Most fundamentally, one is assuming adequate representation, that is, that the presence or response of one or a few indicators accurately reflects changes in habitat quality for an entire community or ecosystem. For example, if northern parula warblers are chosen to indicate the condition of deciduous forests, one is assuming that population levels of this warbler will respond sensitively to the quality of these habitats. Yet this species only occupies a fraction of all deciduous forest habitats, making it a poor indicator for those habitats in which it is scarce or absent. It is therefore preferable to choose enough indicators to represent the full range of habitats present.

Any particular species has specific requirements that differ from those of other species. Thus, northern parula warblers cannot be expected to respond to the same changes in forests to which Nashville or black-and-white warblers respond. Each species will tend to respond uniquely to those particular features of its environment that most affect it. This leads us to a conundrum: if each species responds in its own way to habitat features, then no species is able to adequately characterize the suitability of its habitats for other species (Landres et al. 1988). Again, we are forced to conclude that managers should not rely on too narrow a set of biological indicators, even within one type of habitat.

We also tend to assume that the presence or population density of an indicator accurately reflects habitat conditions. This itself could be misleading in that many species occupy habitats temporarily and some habitats are much more productive than others. For example, northern parula warblers may occur in many habitats but may experience high rates of nest parasitism in some of them. In this case, presence of this species would be a misleading indicator of habitat suitability.

Finally, it also is important to choose indicators that respond quickly to changes in their habitats. Populations of long-lived birds and mammals usually contain adults able to persist even in the face of conditions where successful reproduction is impossible. Monitoring adults in such species may only reveal changes too late to guide management. It would therefore seem advantageous to choose indicators that respond quickly and sensitively to habitat conditions. This, however, can also be taken too far. Some indicators may actually be too volatile, making their population responses difficult to interpret or unreliable as indicators. The potential for rapid population growth in many insects and small mammals, for example, may cause them to undergo wide population fluctuations that may have more to do with temporary climatic conditions or rates of predation or parasitism than habitat quality. Populations with high intrinsic rates of increase may even be prone to chaotic fluctuations which are impossible to predict and only loosely related to habitat conditions (May 1981).

In summary, there are many trade-offs involved in choosing which indicators to monitor. Some are simple to monitor, others provide more valuable information. Rare species may appear preferable as indicators since they often represent what managers want to protect. Their rarity, however, makes them difficult to monitor accurately. In contrast, common and abundant species that are easy to tally tend to be habitat generalists whose populations may only poorly reflect specific habitat conditions. They may also be cosmopolitan or weedy, indications that they are not likely to be adversely affected even by intensive patterns of management. Because resources for monitoring are always limited, managers should avoid common or ubiquitous species as indicators unless they are specifically seeking a negative indicator to reflect disturbed conditions.

Managers and biologists have an understandable inclination to choose conspicuous and abundant diurnal animals as indicators because they are easy to locate and count. Such indicators may restrict detection of some kinds of habitat change. Smaller and more localized vertebrates such as salamanders may be more sensitive to many types of habitat disturbance, while butterflies or other insects may better reflect habitat conditions or the presence and distribution of particular food plants. A number of such specialized indicators will be needed to span the range of habitats present.

What Should We Measure?

Forest managers place great emphasis on the need for accurate and up-to-date data on the productivity of their forests, compiling considerable information on soil types, site index, stand composition, and growth rates. It is logical that they should seek data of comparable quality and quantity regarding the wider set of nontimber species they manage. In this section, we survey the range of ecological indicators to consider in monitoring programs.

Single species approaches

The commonest approach to biological monitoring is to first identify a representative set of indicator species then track their population densities and distribution at regular intervals. Once indicator species have been chosen, field methods are employed to track changes in their distribution and abundance. Conspicuous birds and diurnal mammals are usually tracked using visual counts or vocalizations. For less conspicuous animals, it is usually more practical to gauge densities via counts of scat or other sign. Nocturnal, subterranean, aquatic, or otherwise shy animal species are often trapped, marked, and released for later recapture. The design and analysis of mark and release surveys are highly developed (Ricker 1975; Otis et al. 1978). For game species, it is customary to use data on the age, sex, and condition of the animals killed by hunters to monitor populations.

Birds are often chosen as indicators because they are well-known ecologically and often respond sensitively to habitat conditions. Singing males are counted during the breeding season by listening for set periods at a series of sample points. Each male is assumed to be associated with a territory, yielding a more accurate picture of population viability and growth rate than simple counts of all adults, which would include nonbreeding "floater" individuals. Unfortunately, because unmated males tend to sing more often and later into the season than mated males, these data may still be optimistically biased (J. Faaborg, pers. comm.).

Some have argued that neotropical migrant birds are ill-suited as indicators because their population densities reflect hazards of migration and overwintering mortality in tropical countries experiencing rapid deforestation. However, these species represent a substantial fraction of total bird diversity in eastern forests and tend to be more closely tied to their habitats than nonmigrants. These birds also appear more sensitive to forest fragmentation and other habitat changes in eastern deciduous forests than resident birds (which, interestingly, seems to be less true in the Pacific Northwest—Hansen et al. 1992). It therefore appears inappropriate to discard neotropical migrants as potential indicators, particularly if monitoring can include some measure of nesting success.

Once data are derived from such surveys, how are they used? Most immediately, we gain information on population density and distribution indicating which species are present and in what abundance. Over time, we can also gauge whether populations decline, perhaps signaling deterioration in some key feature of their habitat. Managers should also be concerned when densities of certain species increase dramatically. Invasion of exotic weeds or diseases, for example, often threaten native species (McKnight 1993). Even sustained increases in native species such as deer may spell trouble if they act as a "keystone" herbivore or predator (Garrott et al. 1993).

While population density and distribution provide important information, managers should also seek additional data. Noting the age, size, and gender of the individuals sampled, for example, greatly enhances the value of demographic data by allowing population biologists to construct more realistic models of population growth. This is especially important for long-lived species, where even total reproductive failure may produce only negligible population declines for many years because of the persistence of long-lived adults (the "living dead" phenomenon). Declines in early age classes can provide crucial early warning of imminent population changes in such species.

Spatial analyses are likely to provide additional insights into how forestry practices affect populations. Continuous data will allow biologists to track changes in the range or distribution pattern for species they are monitoring, revealing patterns of invasion, responses to climate change, and responses to local or regional patterns of management. Demographic data within subpopulations may further indicate which are acting as sources and which as sinks. These analyses should help illuminate geographical aspects of population structure and dynamics.

Multispecies and process approaches

We should also extend traditional single-species approaches to monitor multiple species, communities, and ecosystem characteristics or processes (Noss 1990). These are both significant elements of diversity in their own right and can often illuminate critical processes that single-species data might miss. It may also be more efficient to combine data across several species in monitoring programs. For example, a qualified birder can record all the bird species heard at a point almost as quickly he or she can record one or a few species. If most of the time and effort in a plant monitoring survey is spent traveling to the sites and establishing survey transects, sampling most of the plant species present at those sites may be practical.

Data from multiple species also allow more types of analysis and more sensitive tests of some hypothesized trends. For example, managers may be more interested in the relative abundance of two competitors, or a prey and a predator species, than in the absolute abundance of either. Community sampling also allows one to estimate overall diversity and patterns of relative abundance and rarity. Because rare species are intrinsically hard to monitor, it may prove easier to detect their collective presence within reasonably large community samples. It may also prove useful to examine multispecies data using quantitative methods of analysis such as ordination or clustering techniques to compare communities or contrast the effects of alternative systems of management.

In many cases, it may be appropriate to monitor ecological processes rather than particular species, vegetation types, or communities. Ecological processes such as community photosynthesis, respiration, decomposition, growth rates, and productivity are of inherent interest to managers interested in potential yields. They also are essential for assessing the net contributions of young and old forests to the global carbon cycle. It also is of great importance to monitor the frequencies and intensities of various types of disturbance in order to compare managed and unmanaged forests and to gauge the roles these play in maintaining population recruitment and community diversity.

Principles to Guide Research and Monitoring Efforts

To obtain the quantity and quality of information needed to wisely manage biotic resources, land managers will need to foster additional research and monitoring on their lands and cooperate with one another across agencies to obtain and respond to these data. Research is a slow and intensive process, however, and monitoring efforts may require years before clear trends are established. This makes it all the more necessary to plan this work carefully and sustain it through time.

To summarize our points above, we present a set of principles here that we believe should guide research and monitoring. These recommendations parallel those authored by one of us (DMW) for the "Scientific Roundtable on Biological Diversity" (Crow et al. 1993).

1. *Inventory and monitoring efforts should be expanded and systematized to place them on the best scientific footing and ensure a continual yield of high-quality and timely information.*

    Managers of most public forests have surprisingly incomplete knowledge of which species occur on their lands. Typically, information exists only for birds, mammals, trees, and perhaps other vertebrate animal species and rare and sensitive plants. Managers should therefore work to extend the domain of such survey work to include other groups of organisms (cf. Chapter 2). All such work should be carefully planned and conducted regularly and thoroughly enough to ensure statistical validity and biological relevance. Comprehensive sampling designs will be needed to accurately assess population declines, incremental losses of habitat, and changes in ecosystem function. Managers should also seek to link these data more effectively with dynamic models of how species and communities respond to changes in the intensity and configuration of management activities. For example, information on the aerial extent and spatial proximity of stands would allow one to couple existing inventory and monitoring data to dynamic models of likely ecological change.

    Eminent biologists have been arguing for some time that efforts to inventory elements of biodiversity, both nationally and internationally, should be greatly expanded (Campbell and Hammond 1989; Raven and Wilson 1992). The Department of the Interior under current Secretary Bruce Babbitt plans to combine research and monitoring efforts within the U.S. Fish and Wildlife Service and the U.S. Park Service into a unified National Biological Survey (NBS). Serious efforts will be made to map U.S. ecosystems and compile and computerize existing data relative to conservation (Stevens 1993a). While these data will certainly contribute to our abilities to anticipate how development might threaten biodiversity, we should keep in mind how limited our information is, even for the best-explored temperate forest habitats. There are dangers that in the excitement of pursuing a major new program like this, we may tend to forget how little of our total biota is contained in these databases. We may also confuse detailed distribution maps with our far less complete knowledge of how these species will respond to changes in their landscapes. While it is too early to judge the administrative success of this reorganization, it potentially offers a vehicle to achieve a broader scope, better consistency, and more sophistication in biodiversity research and monitoring efforts among our federal agencies. It is critical that parallel investments of similar or greater magnitude be made toward systematic, field-based inventory efforts and ecological research aimed at expanding our knowledge base. While satisfyingly concrete, compilation and computerization of existing data alone will fail to guide us through the difficult management decisions we face in the future.

To ensure consistency and facilitate cooperation among regions, foresters and ecologists should also agree to standardize systems of ecological classification. Sophisticated forest management requires frequent access to the detailed information contained in ecosystem classification systems. To this end, the Forest Service continues to develop its own ecosystem classification system (ECS) to integrate more ecological and geographical factors (Cleland et al. 1992). This information obviously needs to be both extensive and intensive, necessitating a hierarchical approach. The Forest Service ECS system relies on soil, climate, topography, and general vegetation types present in a region to describe regional variation at the "province" and "section" levels. On more local scales, the system extends to include "Land-type Associations (LTA)," "Ecological Land Type (ELT)," and "Ecological Land-Type Phase (ELTP)."

Many classification systems used by forest managers were originally designed to predict the capacity of sites to grow particular timber tree species rather than for analyzing biological relationships. Because the habitat types they recognize are based on soil and site conditions, they often employ species for classification that are relatively insensitive to logging and disturbance. Such systems do not reflect current species composition, handicapping their ability to address biodiversity issues

(Solheim et al. 1991). More biologically oriented systems would therefore emphasize present composition, including site history and successional status. While there is nothing wrong with the use of an ECS, it would be misleading to assume that any classification system, no matter how detailed, could substitute for accurate and timely field data.

2. *Research and monitoring programs should employ the best contemporary scientific knowledge and methodology. To ensure this quality, such programs should undergo periodic peer review.*

Agency personnel should always seek to use limited research and monitoring resources in the most efficient way. To accomplish this, agencies should employ skilled and knowledgeable technical personnel with adequate training in contemporary principles of ecology, population biology, and conservation biology. Where such personnel are in short supply, agencies should consider retraining existing qualified staff or contracting out research and monitoring projects to independent scientists. Research and monitoring efforts should also be routinely reviewed by outside parties. Federal agencies such as the Geological Survey routinely apply peer review to approve research proposals and evaluate results and interpretations. The Forest Service and the Bureau of Land Management should extend their use of formal peer review to include the design, establishment, and revision of biodiversity research and monitoring programs.

3. *Research and monitoring should emphasize those elements of diversity thought to be vulnerable to extirpation, sensitive to man-made disturbances, or keystone species with cascading effects on other elements of diversity.*

Resources are too few to allow comprehensive monitoring of all elements of diversity. Rather than tracking ubiquitous or common species, research and monitoring efforts should emphasize those species and communities known or suspected to be in decline. Special efforts should be dedicated to track species or habitat elements of conservation concern and species or elements known to be vulnerable to widespread anthropogenic disturbance. It may also be important to track particular common or increasing species if these species play a crucial ecological role and so influence community composition or the ability of rarer or more threatened elements of diversity to persist.

4. *Inventory and monitoring efforts should be extended to include other important but obscure groups of organisms.*

Thus far, forest managers have concentrated their biological inventory and monitoring efforts on a few vertebrate indicator species, yet other more obscure groups may make better indicators for many purposes. To ensure that some critical elements of diversity are not being overlooked, additional efforts should be made to inventory alternative taxonomic groups and assess their potential for monitoring. Amphibians, for example, are known to be ecologically sensitive to habitat conditions and are relatively easy to monitor via their calls. They also appear to be experiencing widespread declines.

Some of these relatively obscure taxonomic groups are also known to be functionally important. Lichens and invertebrate animals could provide good indicator species in many instances (Young 1988; Noss 1990). Bacteria and fungi play critical ecological roles as diseases, mycorrhizae symbionts, and decomposers. Lichens and fungi also play key roles in nutrient cycling and frequently fix nitrogen. Given the sometimes dramatic differences in abundance of these groups between young stands and old growth (e.g., Lesica et al. 1991), we should further assess the distribution and functional significance of species restricted to older trunks and forest types. In Europe, many soil fungi have declined for unknown reasons (Jaenike 1991). Clearly these groups merit additional research and monitoring efforts.

Invertebrates represent the highest fraction of forest diversity and are functionally important as well, both as pest species on trees (e.g., the spruce budworm and woolly adelgid) and as biological control agents (the many parasite and parasitoid species). Their generally short life cycles have the virtue of allowing population levels to respond quickly to changes in environmental conditions, but such sensitivity to conditions may also make it difficult to distinguish real environmental change from "noise." Studies of old-growth forests in North Carolina and western Oregon suggest greater relative abundances of herbivorous insects such as aphids in young stands, but greater densities and diversities of predaceous and parasitic invertebrates in old-growth stands (Schowalter and Crossley 1987; Schowalter 1989; Moldenke and Lattin 1990a,b). Such patterns should be investigated elsewhere to assess their generality and functional significance (i.e., how they influence the susceptibility of forests to pest outbreaks). Many arthropods are relatively easy to sample, but identification requires collaboration with experts. Butterflies, however, are easy for amateurs to identify and also include species known to be both rare or threatened and sensitive to ecological function (e.g., the Karner blue butterfly and its dependence on fire-generated habitats to support its food plant, lupine). Such groups may eventually become mainstays for monitoring efforts.

5. *Where possible, use demographic structure or other early warning signs to assess changed ecological conditions rather than simple population numbers.*

It has been customary to monitor populations by their size or density, using declines in abundance to trigger more intensive conservation efforts. Numbers alone, however, often give little information on population viability. Such approaches may provide too little data too late for long-lived species in which adults may persist long after conditions have become untenable for their offspring. In addition, it will often be statistically difficult to demonstrate declines for rare species with limited sample sizes (e.g., breeding bird surveys—Terborgh 1989).

To provide earlier warning and better statistical power, resource managers should monitor demographic variables such as nesting success, seedling establishment, and the size or age structure of populations, particularly for species that are long-lived or of special concern. In addition to providing earlier and more sensitive indicators of imminent population decline, such variables can also often provide insights into the mechanisms driving these population losses. While these methods are often more labor intensive, the data they produce are often of much greater value than simple trends in population size.

6. *Monitoring should occur at a hierarchy of geographical scales.*

In addition to monitoring species, forest managers should gather and interpret data across several geographical scales (Noss 1990). At the highest (landscape) scale, such efforts should include data on forest types, openings, surrounding land uses, the distribution of potential corridors, and measures of habitat area and the degree of fragmentation. Such data are increasingly available via remote sensing techniques and geographical information systems. At the community level, forest managers should assemble consistent information on cover type from the evolving Ecological Classification System (Hoppe 1990) and explore and test the utility of these data for biological monitoring and planning. Data at the species level from regular monitoring efforts could then be integrated into these higher geographical levels to provide better understanding of how management is likely to affect edge-, area-, and isolation-sensitive species.

7. *Inventory and monitoring efforts should include entire guilds or communities in cases where such sampling is efficient.*

As discussed above, sampling many species often provides opportunities for other types of analysis without entailing much additional effort. Data from full guilds or communities allow managers to track more elements and processes than data from single species, providing more statistical power to detect systematic changes (i.e., changes in ratios of abundances or other composite indexes can be used rather than tracking a single time series). Inventory and monitoring efforts should take advantage of such economies of scale whenever possible.

8. *The results of research projects and monitoring efforts should be closely integrated with forest management.*

Because research and monitoring provide information essential for informed management, research and monitoring efforts should be "built in" to ensure that accurate, up-to-date information is available at appropriate points in planning and management cycles. Clear channels and procedures are needed to regularly and reliably inject this information into management. Ad hoc arrangements among forest planners, managers, forest ecologists, wildlife resource specialists, and researchers may not consistently provide opportunities to integrate research and monitoring results with specific projects. Such integration should also tie decisions at each level to diversity concerns at that level.

It also is essential that management respond to new data on an ongoing basis. Such "adaptive management" recognizes that continuing advances in ecological and conservation science will make state-of-the-art approaches today obsolete tomorrow (Walters 1986). Thus, forest managers should allow planning and management to evolve in tandem with our growing scientific understanding of the consequences of those decisions.

In addition to management benefiting from research and monitoring, research and monitoring themselves benefit by being more closely integrated with management. Each management decision provides, in effect, an opportunity to pursue research into the consequences of that decision, provided adequate planning and controls are established. Researchers can embark on larger-scale and more ambitious research projects than are possible solely via research funds. Alternative patterns of cutting, for example, could be integrated into a research design with appropriate controls. Such a system of research and monitoring is particularly important in attempting to restore certain habitat types. Both adequate funding and institutional commitment are needed to achieve this closer integration of research and monitoring with management.

Research Questions

As monitoring programs expand, forest managers will have the opportunity and responsibility to address several research questions the answers to which will bear directly on their management. Such individuals may find it useful to consider (and add to) this list.

1. *What rates of loss, fragmentation and/or degradation are occurring for particular habitat types?*

   Maintaining adequate areas of each distinct habitat type would immediately protect many of the components, structures, and functions of those habitats and therefore serve as an efficient coarse filter to conserve biodiversity. This top-down approach will require comprehensive inventories, consistent systems of community classification, and regional gap analyses.

2. *Are old-growth forests unique? Are there old-growth dependent species?*

   Current management decisions often are based on incomplete or inadequate knowledge regarding the structure, composition, and function of old-growth communities (Franklin 1992; Carey 1989; Maser 1989). Such communities are now quite scarce, especially in the eastern U.S., and have lost top carnivores and other elements of diversity. Nevertheless, they provide inocula for restoration and an essential baseline against which to judge ongoing management and restoration efforts (Martin 1992). We therefore need additional research on the species composition and unique characteristics of these communities, such as soil composition and dynamics (including mycorrhizal associations and ecological roles of soil invertebrates), relationships of herbaceous species to disturbance and soils, and dispersal characteristics and population dynamics of associated vertebrate species (birds, small mammals, bats, reptiles, and amphibians).

3. *To what extent can silvicultural practices or other methods of active management successfully mimic these old-growth characteristics?*

   Given baseline information on old-growth communities, managers should next determine how species typically associated with old growth are affected by active (and passive) methods of forest management. We need research to investigate the degree to which techniques of manipulative management mimic natural processes in old-growth stands and so provide the structural and functional features needed by old-growth-associated species. Work by Franklin and his associates in the Northwest on snags and coarse woody debris suggests that these biological "legacies" provide key habitat characteristics for several old-growth species. Are they enough? Additional work is needed to extend these results to other groups of organisms and other forest community types. Further research is particularly needed to assess the efficacy of artificially producing gaps, tip-up mounds, groundfires, and other types of disturbance in order to provide for the regeneration of disturbance-dependent species.

4. *How do disturbance-sensitive elements of diversity respond to various methods of timber harvest?*

Herbaceous plants, reptiles, amphibians, etc., may respond differently to different patterns and scales of timber harvest. Research is therefore needed to determine how alternative silvicultural treatments (e.g., differently sized patch cuts, different intervals of harvest, "feathered" edges, and leaving behind snags and/or live trees) affect the abundance and persistence of disturbance-sensitive elements. Some of these effects may be indirect, as for the herbaceous plants that depend on ants or mammals for dispersal.

5. *What are the ecological effects of multiple, short-rotation tree cutting? Do such effects threaten ecological or economic sustainability?*

Forest management in the upper Midwest and South is often intensive and based on short-rotation aspen, pine plantations, or other even-aged systems of silviculture. Proposals to alleviate cutting pressures on some areas (Chapter 11) could intensify cutting on other lands in order to meet timber outputs. Given the frequent destructive effects of previous logging (Chapters 1 and 8), we cannot assume that today's intensive forestry will be benign. Thus, we should expand research into the short- and long-term ecological impacts of short-rotation harvests. Here, we are particularly concerned that such harvests, especially when based on whole-tree harvesting, may alter soil structure or deplete essential soil nutrients (e.g., nitrogen and phosphorus). In the upper Midwest, there also is some evidence to suggest that short-rotation harvests of aspen may exacerbate infestations of the fungal pathogen *Armillaria* (Stanosz and Patton 1987). Such serious threats to long-term productivity deserve attention by broadly trained forest scientists.

6. *What is the ecological significance of gap formation?*

Gaps influence a wide variety of species, directly or indirectly, and occur at a variety of sizes and frequencies in natural landscapes (Chapter 4). There is thus the need to assess how various gap sizes and types affect particular components of diversity thought to depend on such disturbances (e.g., tree seedlings and herbaceous plants). Such research should include the effects of microtopography, microclimate, and local light levels on plant growth and recruitment.

7. *How are overharvesting, poaching, and harassment of particular wildlife species affecting the distribution or persistence of these species?*

Many once-common species are routinely collected from public forests (e.g., wild ginseng and club mosses in the upper Midwest). Other species are sensitive to inadvertent human activity (e.g., road density, snowmobile noise) or actively persecuted (e.g., hunters intentionally or unintentionally shooting wolves). Research is needed to assess how wildlife species respond to various degrees of human presence and exploitation. Social science research that addresses why some individuals chase, harass, or shoot nongame wildlife, and how such behavior might be modified, would also be useful.

8. *To what extent do invading exotic species threaten other elements of diversity?*

Invading exotic species such as purple loosestrife, garlic mustard, spotted knapweed, gypsy moth, zebra mussel, wooly adelgid, and other forest pests are conspicuous and often occur in high densities. Because many of these species represent a threat to natural communities, it is important for managers to monitor their spread.

Most of these invading species thrive in open, disturbed habitats and frequently disperse along roadsides or attached to boats or vehicles. Their tie to disturbed sites also makes them useful as "negative" indicators of habitat quality for native species sensitive to anthropogenic disturbance. Because these species are usually conspicuous and abundant after invasion, it should be easy to obtain adequate sample sizes for statistical significance. Monitoring should be coupled with research designed to assess the effects of roads, disturbance, and fragmentation on their spread and the degree to which increases in these indicators threaten sensitive native species.

9. *What ecological processes regulate abundance or opportunities for population recruitment in rare species?*

Concerns for diversity in general and rare species in particular have frequently been limited to identifying a few key species of concern, then monitoring population levels or presumed habitat needs for those species. As we gain a more dynamic view of populations, we recognize that successful recruitment and persistence often depend on disturbance, interactions with herbivores or carnivores, or other key ecological processes. This emphasis on process points up the need for more detailed and dynamic information on population structure and demography.

10. *What are the ecological effects of edge habitats?*

Edge habitats are increasingly widespread in human-dominated land-scapes and threaten several components of diversity by boosting densities of keystone herbivores, small opportunistic predators, and nest parasites (Chapter 5). Edges may also serve as corridors to boost the invasion of ex-otics. Nevertheless, some controversy remains as to the degree to which these processes operate in primarily forested landscapes. To resolve this uncertainty and to assess the severity and scale of these effects, agencies should support research on the degree to which roads, clearcuts, and other natural and human-caused disturbances affect various elements of diversity. For example, we need to assess regional variation in the degree to which neotropical migrant birds are suffering increased rates of nest predation or nest parasitism and the degree to which opportunistic predators and parasites use roads or edges to search for prey. Poaching of game and nongame species also appears to occur most frequently near trails or roads. Although these are obvious examples, there may also be more subtle impacts of roads, trails, or other edges on the distributions or abundances of rare and threatened species.

11. *Do the effects of habitat fragmentation interfere with dispersal or recoloniza-tion in particular species so as to threaten their populations?*

Metapopulation dynamics greatly influence the ability of some popu-lations to persist (Chapter 6). Some subpopulations appear to serve as sources for dispersing individuals, while others clearly act as net popula-tion sinks. Species with limited abilities to disperse across barriers (e.g., some birds, many herbaceous plants, salamanders, etc.) are therefore vul-nerable to the effects of habitat fragmentation. Species whose population dynamics involve repeated dispersal to recolonize sites after disturbances may also respond sensitively to area dynamics. Other species shift their distribution continuously in response to short- and long-term shifts in environmental conditions. Such shifts may accelerate in the future with long-term climate change. Thus, monitoring over long stretches of time will be essential to track the interacting effects of forest type, patch size, fragmentation, and edge on the persistence of sensitive species.

12. *To what degree do rare or threatened elements of diversity depend on dispersal or recolonization via corridors of appropriate habitat?*

Many biologists suspect that habitat corridors play critical roles in the persistence of some area- and disturbance-sensitive species. Unfortun-ately, few experimental studies and data yet exist on this subject, leaving considerable uncertainty about which species need, or benefit the most from, corridors (Hobbs 1992; Simberloff et al. 1992). It is even conceivable that certain types of corridors might threaten some elements of diversity by increasing edge habitat or providing avenues for the spread of pest species. There is thus a need to investigate how effectively strips of habitat function as biological corridors and which species benefit from such "connectedness."

13. *How does forest unit size affect the current and likely future diversity of an area that is temporarily or permanently isolated?*
    To address the critical questions regarding the size of reserved habitat areas which might be needed to adequately protect sensitive elements of diversity (Chapter 11), we need further information on species–area curves for representative plant and animal species groups within the various community types recognized in different regions of the country. Ideally, such research should also involve studies of rates of species loss (relaxation) and immigration or recolonization (turnover) predicted by island biogeography (Chapter 6).

14. *How do high densities of keystone species affect other elements of diversity?*
    Keystone predators and herbivores may directly or indirectly affect many elements of diversity by affecting the relative abundance of competing species. For example, white-tailed deer appear to be acting as keystone browser by driving forests toward a composition dominated by species that tolerate deer browsing (Alverson et al. 1988). At their current high densities, deer may prevent regeneration in important tree species such as eastern hemlock, white pine, and white cedar and decimate populations of herbaceous lilies and orchids. Deer browsing also appears to be having cascading effects on bird and mammal species that depend on understory cover (McShea and Rappole 1992; deCalesta 1994a). Finally, high deer densities may act to exclude other hoofed mammals from the region by providing a reservoir for a parasitic worm that infects the brains of elk and moose (Gilbert 1974).
    Managers need further information on the regional extent and severity of the browsing impacts and more specific knowledge regarding what threshold densities or patterns of distribution damage elements of diversity. Deer densities appear to respond sensitively to vegetative cover and patterns of management. In the upper Midwest, deer increase in the presence of ample aspen and summer grass forage, evergreen forage in winter (logging tops), and the thermal cover provided by conifer stands in winter. Quantifying these relationships will assist forest managers in understanding how their management decisions influence deer abundance. As

these relationships become better known, forest managers should work with wildlife managers to achieve stable populations of deer compatible with sustained forest diversity and productivity.

Beaver may also act as a keystone species because their dams have cascading effects for many other plant and animal species. Beaver dams typically kill surrounding trees and reduce a stream's suitability for some fish species. They also provide habitat for marsh species like ducks that may contribute to regional diversity. Beaver may also serve as important prey for top carnivores like wolves. Forest managers should therefore support research to ascertain how sensitively beaver population levels respond to surrounding habitat management (e.g., aspen) and how beaver influence other local and regional elements of diversity.

Summary

To plan properly, forest managers require accurate knowledge of the likely consequences of their forest management practices. Assessing such impacts for a range of alternative plans represents an important and routine part of the environmental review process. Instead of certainty, however, managers are faced with uncertainty. While scientists often suspect significant threats, they remain uncertain regarding exactly how significant or pervasive the impact is. For example, we still do not know how the loss of natural individual tree gap dynamics across a region may threaten the persistence of particular herbaceous plant and invertebrate species. In other instances, we recognize a threat and have scientific confidence that the threat is real and substantial but do not yet know how effectively some alternative forest management approach might ameliorate the threat. An example would be the degree to which edge-sensitive bird species are responding to roads and timber harvests in northern Wisconsin (Howe et al. 1992). In such cases, scientists recognize the need to monitor the effects of ongoing forest management on particular elements of diversity, both to catch unanticipated declines and to gauge the extent to which management may encourage the recovery of sensitive elements.

To be useful, monitoring data obviously need to be gathered in a consistent fashion over many years to provide a baseline against which to judge change. Thus, there is a need to standardize methods, develop reliable techniques, and adequately train personnel. It is even more important to choose indicator species wisely to ensure that they accurately and sensitively reflect changes in ecological circumstances. Species insensitive to local ecological conditions or species overly sensitive to vagaries of climate and weather are poor choices in that changes in their abundance will be difficult to interpret and unreliable as indicators of the effects of management. Similarly, managers should favor species known to be, or suspected of being, edge-, area-, or isolation-sensitive if they wish to track the ecological effects of forest fragmentation. We combined research and monitoring in this chapter because each obviously benefits from the other: research can inform monitoring, while routine monitoring will provide long-term data essential to answer certain research questions. Thus, research and monitoring will naturally coevolve and should both be considered integral aspects of adaptive management.

Who should do the research and monitoring needed to assess the status of biodiversity in public forests? This is an important policy question, as the Forest Service has been criticized for how it uses science and, occasionally, how it treats its scientists (Westman 1990; Office of Technology Assessment 1992; Lawrence and Murphy 1992; AFSEEE 1993). There has long been a tradition for the Agency to rely on its own expertise and personnel rather than competitive grants or independent contracts to outside scientists to evaluate scientific aspects of management. In addition, the historical division between Operations and Research within the Forest Service, designed to protect researchers from the vicissitudes of funding and policy decisions within individual National Forests and Regions, has tended to further isolate field personnel from contemporary conservation science and its implications for monitoring and management. While Agency scientists and personnel continue to acquire increasing expertise in the area of conservation science, the importance and complexity of these issues justify our recommendation that biodiversity research and monitoring programs be regularly subject to independent peer review. Such reviews, like the conservation and monitoring programs themselves, should be hierarchical, with local, more specific, programs evaluated by scientists with local expertise and the regional and national policies reviewed by prominent scientists. The National Academy of Sciences, for example, regularly establishes panels to review and participate in scientific issues of similar national significance.

The ideal but elaborate set of monitoring needs and research opportunities reviewed here raise serious questions regarding our ability to fully track the impacts of our forest management activities. Forest managers will be particularly reluctant to pursue extensive and intensive monitoring programs when they face reduced budgets and shortages of qualified personnel. How should forest managers manage their forests if they are committed to conserve biodiversity but lack adequate resources to fully assess the consequences of their actions? Under these circumstances it behooves managers to act conservatively by managing lightly and in accordance with the general principles and guidelines outlined in Part II. Such managers should be justifiably reticent about launching ambitious plans to retain diversity by mimicking old growth via silvicultural manipulation or by artificially exchanging individuals among populations to augment presumably interrupted patterns of natural dispersal. Even with the substantial expansion of research and monitoring efforts called for here, forest managers will continue to face uncertainty regarding the full and cumulative consequences of their actions. They will therefore be forced to continue to make decisions based on incomplete data, supplemented by conceptual models. While further research and monitoring will help to resolve points of uncertainty, it would clearly be inappropriate to use the need for more research to delay or avoid making changes in management that conservation biology has already concluded would likely benefit threatened elements of diversity.

PART III

Approaches to Forest Management

HAVING EXPLORED THE ECOLOGICAL MECHA-nisms affecting diversity and our approaches to monitoring the bio-logical health of our forests, it is now time to turn our attention back to the practice of forestry. In this section we are concerned with evaluating the extent to which various approaches to silviculture are compatible with maintaining viable populations of sensitive forest species and the ecological processes they depend on. We have examined the many ways logging differs from natural disturbances such as fire, windstorm, and individual treefall events. Part II explored the many elements of biotic diversity that depend on maintaining ecological conditions at or near their historical values. Also discussed were the ways in which widespread logging and road construction fragment landscapes, threatening many area-, edge-, and isolation-sensitive species. These dependencies reflect long-term evolutionary adaptations and can not be considered manipulable in the short-term. Furthermore, for every species now known to have such a dependence, there are likely to be dozens of additional species with similar, or even more extreme, dependencies. Although most of these species have not yet suffered declines substantial enough to place them on threatened and endangered species lists, changes in their metapopulation dynamics could compromise their long-term persistence.

How can we adjust our practices of forestry to accommodate this new knowledge and ensure that the diversity of our forests will be maintained? To answer this question, we need to pursue several others. First, what are the dominant methods of forest management, how did they develop, and to what degree do they threaten sensitive elements of forest diversity? We pursue answers to these questions in Chapter 8, where we review past and present approaches to forest management. Although enlightened foresters have always striven to sustain the productivity of their forests, economic and social forces have narrowed the focus of forest management on both public and private lands to emphasize the production of particular tree and wildlife species. As a result, certain other species have unwittingly become threatened. Recognition of this fact has been slowed, however, by the common assumption that active management maintains forest health and benefits wildlife. This assumption crops up often in discussions of multiple use and played a dominant role in National Forest planning efforts in the mid-1980s. We therefore examine the Forest Service's notions of multiple use and their recent record of management in Chapter 9.

If conventional patterns of logging threaten certain components of biological diversity, how can we amend forestry practices to protect these elements? We still face uncertainty regarding how different intensities of management affect forest processes and those species dependent on them. Several alternative forestry practices have been proposed in recent years to better protect diversity. We critically review several of these proposals in Chapter 10 in light of the threats to diversity discussed in Part II. We differentiate these alternatives according to both how intensely forest stands are managed locally and how lands excused from timber production are allocated across the landscape. Again, we concentrate our discussion on those aspects of diversity identified as most sensitive to anthropogenic disturbance, not because these are the only elements of diversity to protect, but because they are the most vulnerable to extirpation.

So-called "new" and "excellent" forestry practices promise to ameliorate many deficiencies of intensive forestry. Nevertheless, we question whether they will suffice to protect species threatened by logging and habitat fragmentation. We also ask the practical question: which strategy for protecting sensitive elements of diversity will be the most economically efficient? In Chapter 11 we argue that it will often be most appropriate and efficient to segregate intensively managed areas from areas intended primarily to sustain sensitive elements of diversity. Once we recognize and accept that certain uses should predominate in different areas, we are in a position to more clearly address the trade-offs inherent in multiple-use management. To be more concrete, we present the specific rationale and criteria we used to design Diversity Maintenance Areas for Wisconsin's National Forests. Dominant-use zoning based on principles of economic productivity and conservation biology could efficiently protect diversity in many landscapes.

8

The Evolution of Forest Management

The scientific principles discussed in Part II collectively suggest that the scale and intensity of silviculture across many of our forested landscapes may threaten, now or in the near future, a substantial subset of species native to those forests. Such conclusions alarm many and have led directly to efforts to substantially reform forestry practices on public land (Bryant 1993; Devall 1993). They also appear blatantly to contradict the conventional wisdom that active management categorically enhances natural values and wildlife populations while protecting forests from threats such as fire, disease outbreaks, and wasteful decay (e.g., Coffman 1992; Roberts 1993).

In this chapter, we review developments in forest management over the last century and inquire into the sources and assumptions of this now suspect conventional wisdom. We will argue that although forestry developed a broadened set of conservation concerns over this period, economic and political forces increasingly came to dominate forest management on both private and public lands. As economic concerns have acted to narrow forestry, disagreement has developed over the degree to which highly managed forests can conserve biodiversity. Thus, we are particularly concerned with understanding the origin and implications of the idea that commodity-oriented forest management protects and enhances "wildlife." Such an understanding underpins our analysis of the doctrine of multiple use as applied on our National Forests (Chapters 9 and 12). For those seeking more technical information regarding how geographical and technological developments influenced American forestry, we recommend Williams (1989) and Cronon (1991).

Stages in Forest Management

To frame our discussion, we distinguish six overlapping stages in the development of forest management in the U.S. We distinguish these stages in order to identify the historical bases for the current stance of forest managers and public agencies charged with protecting our forests. Because we have already described the first stage in Chapter 1 and will focus on the sixth stage in Chapter 11, we focus our discussion here on stages 2 through 5.

1. *Rapacious.* With abundant forests and few legal restrictions, the lumber baron era was dominated by a "cut-and-take" ethos that afforded little concern for environmental values or the sustainability of forest growth. Lands often were abandoned after cutting, slash fires raged, and no reseeding or other active efforts at regeneration were made. This stage extended into the early twentieth century in many areas (see Chapter 1).

2. *Conservation management.* Forestry developed in the U.S. in response to the excesses of the rapacious period. Forestry schools and state and federal forestry programs emphasized soil and water conservation, controlling fires, and regenerating forests on cut-over land. These concerns also sparked the acquisition of "Forest Preserves" for conservation that later became the National Forests. This stage extended from the late nineteenth century through the early twentieth century.

3. *Broadened conservation.* As forestry developed, its concerns grew to encompass a wider set of concerns, including recreation, wilderness, and wildlife population management. This stage, which began roughly in the 1920s and extended through the 1950s, was paralleled and led by Aldo Leopold's evolving concepts of conservation.

4. *Maximum output.* As forestry became professionalized, foresters naturally tended to focus on economic criteria and maximizing timber output. Foresters began to convert forests to more productive species, select tree populations for greater growth, shorten rotations, and manage even-aged stands for maximum production. These forces came to private lands at the start of the twentieth century and spread to public lands by the 1940s and 1950s.

5. *Traditional multiple use.* As production forestry spread to public forests and as game management increasingly influenced public land management, there developed increasing pressure to interpret consequences of more intensive silviculture as beneficial to wildlife and recreation concerns. Such associations allowed foresters to defend intensive management on the grounds that they were simultaneously serving a wide variety of uses. This paradigm emerged forcefully in the 1950s and persists to the present.

6. *Dominant use zoning.* We anticipate that better understanding of the ecological effects of intensive and extensive forest cutting will combine with greater concern for economic efficiency to favor the zoning of forest

lands so as to segregate conflicting uses. Although intermittently proposed since the 1920s, such zoning has yet to be implemented on a wide scale. We anticipate, however, that such zoning will emerge as principles of conservation biology become better known, as public concerns for diversity continue to grow, and as fiscal constraints limit logging in the 1990s and beyond (Chapters 11 and 14).

While these stages are clearly artificial and each stage remains manifest, they provide a framework for interpreting how history has influenced our expectations regarding forest management. In discussing them in turn, we hope to illuminate the sources of our current ideas and regulations affecting forest management. In many cases, we feel that these perceptions present obstacles that have delayed our ability to understand, accept, and apply the principles of conservation biology in forest management.

Conservation Management

From its inception as a profession, forestry sought to reform the blatantly destructive effects of large-scale cut-and-run logging. The devastating effects of deforestation first became evident in the forests around the Mediterranean in Egypt, the Near East (e.g., the cedars of Lebanon), Greece, and Italy (Perlin 1989). As the trees were cut for firewood, charcoal, and lumber, and as grazing by domesticated animals increased, soils eroded from the hillsides and mountains of the region. Plato, centuries before Christ, recognized these connections and decried the loss:

> What now remains compared with what then existed is like the skeleton of a sick man, all the fat and soft earth having wasted away, and only the bare framework of the land being left. (Quote from Ehrlich and Ehrlich 1980)

Recent evidence from Central America suggests that agriculture and logging took their toll in pre-Columbian America as well (Denevan 1992). Historians have long attributed the collapse of many early civilizations in part to the massive ecological effects of deforestation (Marsh 1864; Perlin 1989).

Fires, floods, and other disasters also occurred in the U.S. in the wake of the destructive logging practices of the nineteenth century (Chapter 1). Massive clearing of forests eliminated root biomass and live and dead organic matter in the soil, which act to retard runoff and erosion. Under such conditions, spring rains can have devastating effects, as they did May 31, 1889, in Johnstown, Pennsylvania, when a massive flood killed 2200. The Forest Reserve Act of 1891 and the Organic Act of 1897 were designed to stem the abuses of exploitative logging practices and protect the ability of watersheds to sustain the continuous water flows needed to support shipping. At that time, there also was a growing perception that the country's huge timber supplies were limited and running out in many areas.

The origins of forestry as a profession and the U.S. Forest Service may be traced in large part to reactions against blatant forest damage and the accompanying environmental scars (Steen 1976, 1984; Fox 1981). German immigrants in the late 1800s like Carl Shurz and Bernard Fernow brought with them European forestry concepts that were soon incorporated into schools of forestry (Rodgers 1968). Franklin Hough in 1873 presented a paper to the annual meeting of the American Association for the Advancement of Science citing historical losses and expressing strong concerns over stream flow, watershed protection, and forest depletion (Allen and Sharpe 1960). After lobbying Congress for two years to appropriate funds to study and report on woodlands, Hough was appointed Chief of the newly established Division of Forestry in 1876 (Steen 1976). The forest survey that followed documented egregious impacts of the lumber baron era and led directly to the establishment of the first national Forest Preserves by Presidential proclamation in 1891. As designations of lands for Parks and National Forests accelerated under Teddy Roosevelt, the timber industry recognized that it would not have access to cheap federal lands in the West as they had in the Midwest. This prompted companies like Weyerhaeuser at the turn of the century to purchase large tracts from the railroads, lands that subsequently became the private industrial forests of the Northwest.

With visions of denuded lands, talk of a timber famine, and memories of massive floods still fresh, Congress drafted the Organic Act in 1897, which limited cutting on the National Forest Preserves to "dead, matured, or large growth" trees that were individually marked. Although Gifford Pinchot, the first Chief of the Forest Service, promoted a utilitarian doctrine committed to providing forest resources for human use, the first job the young Agency faced was protecting and restoring forests. Foresters had learned the importance of retaining forest structure during and after logging, especially on the steep terrain that characterized western forests. The first national foresters in the western states were employed mostly to reforest logged areas in order to protect watersheds from erosion, limit grazing, and extinguish forest fires. After the Weeks Act in 1911, the federal government began to buy up lands in the eastern U.S., launching eastern National Forests. These forest lands presented a far different management challenge than the western public forests. In the East, foresters were preoccupied with what we would now term reclamation and restoration, an attempt to reestablish forest cover across broad areas that had been heavily cut, often burned intensively, and subsequently abandoned as they become tax-delinquent. On these lands, fire control and speedy replanting were deemed most critical for reforestation. Thus, the federal forester's job was primarily a custodial one during the initial decades of the twentieth century.

Concerns with the destructive effects of logging were not confined to the public lands. Discussion to regulate forestry on private lands began about 1917, gained momentum in the 1920s, and persisted through the late 1940s (Siegel 1990). Had the National Industrial Reform Act of 1933 not been ruled unconstitutional, it would have regulated private forestry to some degree. National Forest Practice Acts to establish forest practice committees and limit clearcutting were introduced in 1941, 1947, 1949, and 1971, but all of these efforts at federal regulation failed enactment (G. Robinson 1988).

These failures at the federal level led several states eventually to enact their own laws to control logging practices on private lands. These vary widely in the standards they specify and achieve, but all demand regeneration via leaving seed trees or planting when natural regeneration fails. Many are more specific, restricting logging from steeper slopes and near streams or prescribing the distribution and type of logging practices. Washington and some eastern states also target concerns for scenic beauty or amenity values. Recently, as pressure has grown in many states (e.g., Massachusetts) to enact more stringent local county or township regulation, the timber industry has moved to join state Departments of Natural Resources (DNRs) in drafting less restrictive uniform state laws. Although many state forestry statutes seek to protect watersheds and environmental values, none to our knowledge directly addresses the diversity issue. Michigan, however, passed a separate Biological Diversity Conservation Act in 1992 (House bill 4719) that, although nonprescriptive regarding forestry, did win overwhelming support. In addition, the DNRs of many states are now reformulating "best management practices" intended to protect water quality and revising other forestry policies to improve protection of biodiversity and old growth.

Broadened Conservation

As forestry developed as a profession through the early years of this century, it generated a broadened set of concerns. Once foresters had addressed soil conservation and reforestation, they found it natural to extend their concern to conserving wildlife and opportunities for backcountry recreation. Wildlife concerns also emerged because of the widespread loss of big-game species and growing interest in outdoor recreation.

Aldo Leopold's career paralleled these broadened concerns within forestry

(Meine 1988). His first job as a forester in New Mexico and Arizona brought him into direct contact with issues such as grazing intensity, predator control, and protecting watersheds and stream banks from soil erosion. With his friend and fellow Forest Service officer Bob Marshall, Leopold forged ideas and plans for wilderness that included both its special and aesthetic role in outdoor recreation and its role as a standard against which scientists could compare more managed lands (Meine 1988). In general, such designations posed few conflicts since these lands could not be efficiently grazed or logged.

Later in the 1930s, Leopold became concerned with how forestry was influencing the landscape. At the time, foresters were beginning to favor replacing original forests with highly managed stands (see Maximum Output, below). Such "type conversion" had long been the norm in Germany. Mid-elevation forests that had been in hardwoods in the Harz and Black mountain regions had been planted to conifer plantations as early as the late eighteenth century. However, it was eventually noticed (again first in Germany) that the litter from conifers tends to acidify soils, lowering their nutrient status and ultimately their productivity. After two or three generations of softwoods, these effects progressively limit growth and may lead to irreversible changes in soil structure and composition (and biodiversity). The realization that plantation forestry and short rotations can harm forest soils and limit productivity led to the Dauerwald, or permanent forest, movement in Germany (Leopold 1936). Ironically, this occurred just as American foresters were adopting type conversion with increasing frequency and touting its many benefits. Dauerwald emphasized original forest tree types (i.e., deciduous hardwoods) and selective cutting on a sustained yield basis.

Leopold was particularly concerned that intensive forestry and intensive game management were becoming incompatible with each other and with healthy populations of wild plants and animals. When he visited Germany and Austria in 1935, he became alarmed at how artificial feeding and other management techniques were being used to maintain artificially high densities of roe and red deer. Their high densities, in turn, forced foresters to regenerate forest stands by excluding deer with fences (see Heart's Content, Chapter 3). In a critical but little cited paper, Leopold (1936) interpreted the German situation to be a perversion of both forestry and wildlife ecology with many lessons for America:

> . . . we have traced through a period of nine centuries the slow but inexorable growth of a system of silviculture incompatible with a natural and healthy game stand, and of a system of game management incompatible with a natural and healthy silviculture.

Leopold went on to observe that European yew, a source of wood for long-bows and of great economic value for many centuries, had become quite rare and that "The combination of wood-factory silviculture and deer has been ruinous to many species of songbirds and flowers." Years ahead of his time, Leopold recognized the dangers accompanying single-minded forest management and fingered the conceptual error he saw behind this approach:

> [There is an] uncritical assumption, dying but not yet dead in America, that the practice of forestry in and of itself, regardless of what kind or how much, promotes the welfare of wildlife.

To avert such a disastrous outcome in American forests, Leopold made three bold prescriptions:

> First, I would say that a generous proportion of each forest must be devoted entirely to floral and faunal conservation. The National Resources Board has suggested 3 per cent. This seems to me much too small. After seeing German silviculture in practice, I have the impression we would do well to develop an intensive silviculture on the better half of our forest soils and leave the other half primarily to other uses.
>
> Secondly, I would point out that the German impasse is probably exaggerated by the fact that deer predators are extinct. . . .
>
> A third and obvious lesson is a deep respect for natural mixtures, and deep suspicion of large pure blocks of any species, especially species not indigenous to the locality.

With customary prescience, Leopold anticipated three of today's biggest forest conservation issues. Sadly, his early warning and sage advice were largely ignored by his fellow foresters and the Agency that launched his career (Chapter 12).

Leopold became increasingly concerned with ecology and the concept of "land health," by which he meant the balanced interaction among all native elements of the biota (Meine 1988). He abandoned accepted notions of top carnivores as "varmints" and worked to establish evidence that wolves and cougars could help sustain forest health and diversity by limiting densities of browsers like deer that otherwise tend to outstrip their carrying capacity. Unfortunately, his recommendations regarding the need to improve limitations on deer numbers in the 1940s were scorned and even ridiculed in his home state of Wisconsin. Although his positions were often unpopular, his articulate voice and concern for science helped to persuade many to consider a wider set of issues in forest management.

Maximum Output

Forestry and foresters have traditionally enjoyed an enviable reputation. In the halcyon days of logging lumberjacks were romantically glorified as rugged individualists. As forestry developed in the early twentieth century, foresters were respected as well-trained and highly motivated caretakers of their forests, always ready to fight fires or defend forests from other threats. Foresters themselves are frequently drawn to their profession by these images and work to maintain them via public relations efforts emphasizing their sincere interest in protecting healthy forests from harm. Among 13-year olds in Germany in 1991, no other future profession is as popular.

Despite these images, foresters since World War II have largely concerned themselves with economic criteria of forest performance. North Americans inherited many of their forestry traditions from Europe. Gifford Pinchot, trained in France, became the first Chief of the U.S. Forest Service in 1905 and architect of its guiding policies (Steen 1976). Like the European foresters he emulated, he believed in actively managing forests to provide a continuing supply of forest products. As in the U.S., European forestry practices evolved in response to crises of supply. Germany had exhausted most of its forests by the seventeenth and eighteenth centuries, sparking an interest in the importance of replanting and controlling rates of cutting by the 1780s (G. Robinson 1988). Early European foresters logically focused on growing stands of trees for commercial production, favoring conifers for their wood quality and high growth rates. Active steps were taken to minimize disease, breakage during harvest, and other losses. Under such management, seedlings are often planted, stands are regularly thinned to enhance the growth of remaining trees, and trees are harvested at economic maturity—the point where growth rates start to slow. Economic criteria emphasize costs of production, yield, and growth rates. Intensive silviculture succeeds by these criteria: conifer plantations often produce three to seven times as much wood per unit area per unit time as unmanaged stands (Farnum et al. 1983).

Through most of the twentieth century, forestry has addressed with precision the question of how to maximize timber yields, a well-defined goal of obvious interest. As the forest industry turned to acquire and manage their own lands in the early twentieth century, they began to hire professional foresters trained at the nation's forestry schools. This had the natural effect of influencing their curricula to emphasize site productivity, optimum planting designs, and factors contributing to economic return (Hirt 1991). Newly trained foresters, eager to apply their knowledge, began converting more forests to even-aged stands, shortening rotations, and planting selected stock. Indeed, any forester not practicing such "improved" methods risked being labeled old-fashioned or incompetent.

Such intensive approaches reflect a natural orientation for industrial forest lands but soon spread to affect public forests as well. Foresters took advantage of the opportunity provided by logging to plant conifers, favored for their rapid growth and the production of wood and pulp. Thus, Civilian Conservation Corps crews were used throughout the 1930s in National and State Forests in the upper Midwest to plant red or jack pines, replacing stands of mixed deciduous forest with conifer plantations. Clearcutting, first exploited in the West, became increasingly common in public forests throughout the country by the 1950s and 1960s. Such techniques appeared well-suited to meeting a growing country's demand for more timber. The result of these trends was a convergence in goals and management practices between private and public forest managers.

Traditional Multiple Use

The increasingly economic focus of many foresters coupled with increased demands for lumber after World War II brought dramatic increases in the amount of logging on public lands during the 1950s and 1960s (Fig. 8-1). Such increases could hardly escape the notice of those concerned with public lands. As public forest managers increased cutting, shortened rotations, and converted forest types, they came under increased criticism from conservationists who considered these intensive types of forestry to be unaesthetic, destructive of natural values, or otherwise inappropriate (Fox 1981). This group included citizens who were increasingly using their new cars and roads to visit the public forests to camp, hike, or pursue other recreational activities. Those who had imagined that public forest managers served primarily as stewards of the land were offended to see a proliferation of logging roads and clearcutting. Such citizens soon began to decry cutting patterns they saw as destructive or ugly and to lobby for the creation of permanently protected wilderness.

In response to growing criticism associated with increased timber harvests, Congress passed the Multiple-Use Sustained-Yield Act (MUSY) of 1960, which explicitly recognized the legitimacy of wildlife and recreational values in public forest management and required that forests be cropped in such a way that yields would not decrease over time. The Wilderness Act followed soon thereafter in 1964. These laws clearly signaled to forest managers the public's changing expectations and served notice that their activities would face increasing scrutiny. In response, many forest managers felt the need to justify timber harvest activities more assertively to a public they felt were misinformed. In addressing what many considered to be a public relations problem, these managers were eager to seize upon ideas or slogans that re-

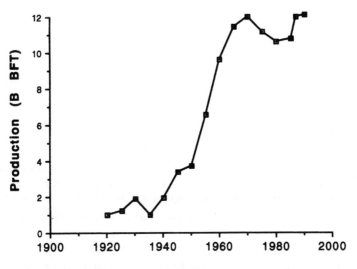

Fig. 8-1. Growth in timber production on the National Forests from 1920 to 1989 in billions of board feet. Note rapid increase from World War II to about 1972. (From USDA Agricultural Yearbooks.)

flected their own perspective and beliefs regarding the positive benefits of active forestry. In particular, they felt the need to demonstrate how forests could simultaneously meet multiple uses while being managed according to conventional silvicultural methods.

Such justifications came easily. While perhaps aesthetically offensive, foresters could argue that clearcuts were necessary to "rejuvenate" the forest, avoid insect infestations, and achieve better economic returns. In contrast, old growth was cast as decadent, over-mature, and prone to wasteful blowdowns or disastrous outbreaks of disease and insect pests. In addition, building roads facilitated access to public forests, boosting recreational visitor days while also providing access for logging operations. These justifications persist. Increased road building was justified in many National Forest Plans in the 1980s on the grounds that they contribute to recreational use (O'Toole 1988; Sample 1990, 1991). Similarly, real or imagined threats to forest health are increasingly used to justify immediate "salvage logging" because such logging is immune to environmental review and avoids timber volume targets that otherwise constrain logging (Keene 1993).

Forest managers began to work more closely with wildlife managers as the scope of conservation broadened during the 1940s and 1950s. As they cooperatively planned timber harvests in the public forests, it was logical for both forester and wildlife manager to add "enhancement of wildlife habitat" to their list of the positive effects of forest management. To understand this justification, one needs to understand what the term "wildlife" implies. As noted in Chapter 5, wildlife biologists beginning with Aldo Leopold recognized that populations of game species such as grouse and deer usually increase in areas where habitats adjoin one another and where plentiful early successional patches are present to provide forage and browse. Edges, like other ecotones, also represent areas of locally enhanced (alpha) diversity. Thus, managing habitats to increase the abundance of early successional tree species, such as aspen, and the amount of edge also tend to favor game species and local diversity (Leopold 1933; McCaffery 1986). Not surprisingly, this has become a focus of many game managers. In fact, habitat management, combined with strict hunting regulations and the loss of most top predators, has been so successful that many biologists now consider densities of deer to be biologically harmful and a threat to sensitive elements of diversity (Alverson et al. 1988; Garrott et al. 1993; Chapter 5). Private industries also recognize the public relations opportunity here. A two-page spread by the Southern Forest Products Association in the April 1973 issue of *Natural History* magazine proclaimed that "Planned forests . . . helped increase the deer population in the South by 800% since 1940" and elaborated to explain:

> Part of the reason is the periodic thinning of the forest, involving the removal of mature and defective or inferior trees to make room for a healthy growth of superior trees, which creates browsing material and nourishment for wildlife.

Thus, a marriage of convenience resulted between forest managers concerned primarily with silviculture and wildlife managers interested in boosting populations of game species. The idea that intensive patterns of forest harvest actually benefit wildlife and enhance diversity reinforced the conviction of many forest managers that such management enhanced multiple forest values. Foresters appeared to have discovered how to satisfy many uses simultaneously and in a way that dovetailed well with their traditional training in maximizing economic productivity. For forest managers to have their cake and eat it too, it was only necessary that game be equated with wildlife and local (alpha) diversity be considered preeminent over overall (gamma) diversity. Although almost invisible to the general public, this semantic shift has had momentous implications for the conduct of forestry on our public lands. It also contrasts sharply with Leopold's frank admissions of the impacts inherent in intensive forestry and his admonitions to set aside large areas to protect natural ecological processes.

While it is true that active management boosts populations of many game species, equating game species with wildlife, edge with diversity, and more with better has burdened many foresters and naive outside observers with a seriously distorted view of the relationship between silviculture and biodiversity. The idea that forests can be simultaneously and successfully managed for intensive timber production as well as plentiful and diverse wildlife has hardened into dogma in some circles. Many now construe the Forest Service's multiple-use mandate to mean that these forests can, and should, be managed primarily for timber production and conventional wildlife values. This interpretation has helped to extend intensive management through most of our public forest lands.

Is Active Management Compatible with Conserving Diversity?

To what degree are intensive modes of silviculture compatible with maintaining diversity? While most admit that many species may not thrive in cut-over lands or early successional habitats, many have tried to claim that few or no species depend on extensive tracts of old-growth forest. For example, top carnivores such as the timber wolf appear sensitive to disturbance and road density but can persist in managed landscapes if contact with humans is minimized (Chapter 12). Similarly, some have argued that although neotropical migrant songbirds require particular habitats and suffer high rates of nest parasitism and nest predation in woodlots within agricultural regions, they also often adapt to early successional stands and experience good nest success within primarily forested regions.

While many species appear adaptable and often persist in managed landscapes, it would be a mistake to accept intensive management as benign until proved detrimental given the many mechanisms threatening diversity discussed in Part II and our continuing ignorance regarding unknown elements of diversity raised in Chapter 2. Aside from these general concerns, biologists also recognize a set of species distinctly associated with older forests. Birds like the northern spotted owl, the northern goshawk, and the red-cockaded woodpecker have gained notoriety in public debates over the nation's remaining old-growth forests in the Northwest and Southeast. These charismatic symbols for old growth have also become lightning rods for public debate, as they appear to directly contradict the dogma that logging benefits wildlife. Thus, it is not surprising that spokesmen for the timber industry publicly questioned whether northern spotted owls actually need old-growth forest. The Forest Service's 1984 Final Regional Guide and EIS for the Pacific Northwest Region reflected similar sentiments by allocating only 405 hectares (1000 acres) of old-growth forest per pair of owls to a total of 375 pairs (Simberloff 1987). In reviewing that recommendation, the advisory panel of ornithologists assembled by the Audubon Society recommended that at least 1500 pairs be maintained on protected sites covering 567–1821 hectares (1400–4500 acres) of mature forest. In response, the 1986 Draft EIS Supplement proposed 550 areas of 405–891 hectares (1000–2200 acres) in its preferred alternative.

Discussion also continues over whether active management for particular old-growth associated species could effectively substitute for retaining old-growth forests. For example, Lennartz and Lancia (1989), in a paper subtitled "Opportunities for Creative Silviculture," argue that the red-cockaded woodpecker may only depend on particular features of old-growth forests which could be provided through active management: "Total needs may be met if the selected element and need can be identified and integrated with the management of secondary forests." While this may be true in some instances, it is unrealistic to believe that we will even be able to identify the majority of forest species in the foreseeable future (Chapter 2), much less gather enough information to manage each correctly. Such species-by-species approaches, although also pervasive in our implementation of the Endangered Species Act (Chapter 12), appear both inefficient and prone to failure in attempting to conserve entire ecological communities.

How Sustainable Are Current Systems of Silviculture?

Silviculture today often addresses particular problems one-by-one using intensive methods. If natural regeneration is sporadic, discing or chaining is used to prepare a mineral soil seedbed, seedlings are artificially planted, herbicides are applied to control broadleaf competitors, and antideer tubes or fencing are erected to prevent browsing. If site fertility declines, fertilizers are added. When insect outbreaks threaten, pesticides are applied or a prophylactic salvage cut is ordered to maintain forest health.

Forestry developed as a profession in both Europe and the U.S. in reaction to the hazards of early cut-and-run logging practices. While many assume we have learned our lessons and that current silvicultural methods and standards are sufficient to protect the productivity of our forest lands, there is some evidence to suggest this trust may be misplaced (DeBell 1990). As silviculture comes to resemble agriculture, many have begun to express concerns with long-term soil fertility, losses of biodiversity, and other cumulative effects. Although we are primarily concerned in this book with the effects of past and current forestry practices on biodiversity, we pause here to consider questions regarding other aspects of sustainability.

Surprisingly little work has been done regarding the long-term sustainability of conventional silvicultural practices, perhaps because such studies are intrinsically long-term and broad in scope. Recent studies, however, suggest that some forest lands may suffer reductions in soil fertility or productivity after intensive or repeated harvesting. Modern logging practices vary in their impacts but can result in substantial soil erosion or losses of nutrients that exceed rates of natural replenishment. Short-rotation harvests, like intensive farming, can deplete the soil of critical nutrients, including calcium, magnesium, and potassium, especially if whole trees are harvested (Pöyry 1992). As nutrients are depleted or oxidized to other forms, soil chemistry changes, possibly affecting interactions between bacteria and fungi.

Clearcuts can have major impacts on soil and understory plant and animal communities because they change the environment so drastically. The loss of cover and the disintegration of fine roots leave soils exposed, vulnerable to rain splash, wash erosion, higher temperatures, and oxidation. These processes may accelerate nutrient loss and soil instability. Slash composed of branches, twigs, and leaves is composed of a majority of the nutrients contained in trees. Removing it therefore accentuates nutrient depletion (Bormann et al. 1968; Sollins and McCorison 1981). While losses from soil erosion and leaching vary greatly among sites in response to differences in slope, logging practices, rainfall, and soil characteristics, they can often exceed rates of renewal, especially when bark or whole trees are removed from the site (Pöyry 1992).

Mechanized systems of logging substitute heavy equipment such as "fellerbuncher" rigs, skidders, and big logging trucks for the teams of lumberjacks and draft animals used a century ago. Not surprisingly, such heavy equipment scrapes and compacts the soil as logs are dragged from the forest. Soil compaction was predicted to be a significant impact for about 13% of harvested forest areas in Minnesota in a recent EIS (Pöyry 1992). Mechanized logging also requires an extensive network of logging roads, which are always compacted and present additional hazards regarding soil erosion, particularly on steep terrain.

Issues surrounding global warming have made the relationship between forests and atmospheric carbon dioxide levels a political football. Industry spokesmen, forestry texts, and even politicians (Sen. Stevens, AL—*Anchorage Daily News*, Sept. 21, 1988) have argued that cutting "decadent" old-growth forests and replacing them with faster-growing and more intensively managed forests will retard the increase in atmospheric carbon dioxide by providing better carbon sinks. Young forests do have high rates of primary productivity, and it seems plain that planting a forest where none existed before would represent a sink. Nevertheless, there is little empirical evidence that younger forests are better than older forests in sequestering carbon. In fact, it appears that old-growth stands efficiently store huge quantities of carbon, much of which is released either shortly after cutting or later when wood products decay. The oxidation of slash and soil organics rapidly converts stored carbon to gaseous carbon dioxide (Bormann and Likens 1979b; Bormann et al. 1974; Houghton and Woodwell 1989). The overall effect of logging old growth forests therefore depends both on how great this release is and on how long wood products are expected to last. Harmon et al. (1990) estimate that the loss of old-growth forests in western Oregon and Washington over the past 100 years has released 1.5 to 1.8 billion metric tons of carbon to the atmosphere. Such massive releases of carbon from these and all the other logged primary forests have therefore contributed significantly to the observed increase in atmospheric carbon dioxide.

Biological interactions may also be altered by widespread intensive forest management, with repercussions for forestry. For example, replacing hard-wood stands with conifers can lead to permanent changes in soil chemistry as acidic leaf litter accumulates, reducing soil nutrients and productivity. In the upper Midwest, research suggests that repeated harvesting of aspen via cop-picing after short rotations exacerbates infestations of the fungal pathogen *Armillaria* (Stanosz and Patton 1987). Because so much of the landscape is actively managed for this species, this biotic interaction could seriously com-promise timber outputs. Outbreaks of gypsy moths and tent caterpillars may also increase in landscapes with more edge (Bellinger et al. 1989; Roland 1993).

While foresters continue to pride themselves on their successes, this pride should not be allowed to ripen into hubris or blind us into assuming sustain-ability instead of viewing it as an elusive goal to be pursued. Indeed, the his-tory of natural resource extraction is littered with many more failures than successes, even when intentions were good and attempts were made to in-clude scientific analysis (Ludwig et al. 1993). Further research is clearly needed to assess how soil conditions and interactions with pathogens and herbivores respond to changes in forest cover, techniques of logging, and short rotations.

Summary

Although forestry in the U.S. has emphasized conservation since it developed as a profession at the turn of the century, historical forces conspired to narrow the focus of both public and private forest managers after World War II. As they sought to expand productivity and meet demand for multiple uses, public forest managers tended to emphasize a single style of forestry over most of their lands that they felt could simultaneously maintain forest health, enhance wildlife populations, and produce a continuing supply of timber. While seductive and valid in a restricted sense, the idea that logging benefits wildlife has obscured the real threats faced by sensitive elements of diversity in heavily modified habitats. Furthermore, it has been used to help justify dramatic increases in the intensity of timber harvests on public lands. The U.S. Forest Service, in particular, has chosen to interpret and apply concepts of multiple use that disguise the true environmental costs of their accelerated timber harvest programs and the degree to which alternative uses conflict. We therefore turn to consider this agency and its interpretation of multiple use in more detail.

9

Multiple Use on National Forest Lands

The concept of multiple use pervades the law, culture, public relations, and practice of modern forest management. Perhaps because it has so many facets, the term can mean many different things in different contexts and to different people. Although the elements of multiple use are quite elusive, the idea of multiple use is so central to forest management in the latter half of the twentieth century that any proposals to steer forest management toward a new treatment of biodiversity must address whether and how those proposals conform to or deviate from multiple use. For many, the validity or usefulness of any new models (Chapters 10 and 11) will be measured against the yardstick of multiple use.

The Multiple-Use Sustained-Yield Act of 1960 (MUSY) first codified multiple use as the official land management policy of the U.S. Forest Service. MUSY defines multiple use as

> The management of all the various renewable resources of the National Forests so that they are utilized in the combination that will best meet the needs of the American people; making the most judicious use of the land for some or all of these resources or related services. . . ; that some of the land will be used for less than all of the resources; and harmonious and coordinated management of the various resources, each with the other, without impairment of the productivity of the land. . . . (16 U.S.C. § 531(a))

Valid purposes of the national forests under this definition include "outdoor recreation, range, timber, watershed, and wildlife and fish purposes" (16 U.S.C. § 528).

The definition of multiple use contains no specific goals for any specific resources, and only the most vague of environmental protection standards ("without impairment of the long term productivity of the land"). Notably, the core of the statutory definition is that the mix of activities must satisfy the general public's notion of an appropriate combination. Very few federal agencies are guided by a fundamental principle which is so ambiguous and devoid of explicit substantive standards. Can one imagine the U.S. EPA having its central mission statement being simply "whatever the public wants"? The Ninth Circuit Court of Appeals has stated that the general language of MUSY "breathe[s] discretion at every pore" [*Perkins v. Bergland,* 608 F.2d 803, 806 (9th Cir. 1979)].

Multiple Use and Dominant Use

Under such a vague statutory definition, one can imagine a very wide range of multiple-use management scenarios. Should all uses occur simultaneously in a particular location in order to satisfy the definition? Would it suffice if uses were alternated in that same location, so that over 40 or 80 years a location served well each purpose? Or would it be permissible if within a ranger district or national forest all uses were combined in some way satisfactory to the public even though in any particular 40-acre patch, only one use was dominant, and others virtually excluded, for a human life span? May each of these combinations be considered multiple use?

If the concept of multiple use means that all uses should occur simultaneously, one might envision a sylvan scene in which two hikers stop to photograph a wildflower while a logging truck slows down for browsing deer; and just downslope and behind a huge bole left by the loggers to provide salamander habitat, trout flash in the glassy pools of a clear brook. At the other extreme, one could imagine an area from which all native vegetation had been removed and replaced with a field of alfalfa, which is cropped periodically for livestock feed and replaced within the next decade with a plantation of walnut trees. Thus, in the long run, the alfalfa field and the walnut plantation are on land that could be considered managed as multiple-use land.

More commonly, multiple use is exemplified by an area which is clearcut today (timber), used as browse for deer in another decade (wildlife, recreation), and the site of skiing or snowmobiling trails in two or three decades. Certain uses predominate for short periods; but over even a period as short as a human life span, many uses may occur in the same acreage. As uses are preferred by humans and animals over time and over the relevant region of concern, the humans and animals move about the forest responding to changing local conditions. At any snapshot in time, many if not all conceivable uses are ongoing at several points in the regional landscape.

Such rearrangements of uses about the forest are a result of the fact that some uses are not compatible with other uses which might otherwise occupy the same space and time (Robinson 1975). Do these incompatibilities require that there be but a single use at any one spatial and temporal point? The often heard accusation "single-use land" cannot be interpreted literally; it refers instead to dominant uses (such as wilderness) viewed by some groups as un-desirable uses of forest land. Economist Marion Clawson reached a similar conclusion nearly two decades ago:

> If single use is interpreted to mean that one use, and only one use, is made of a forest to the exclusion of all other uses, then this situa-tion is nearly always impossible. Some functions exist to some de-gree on all forest land, whether by neglect or in opposition to man-agement. (Clawson 1975, p. 41)

Wilderness serves purposes, at a minimum, for watershed, recreation and wildlife and fish, and intensively harvested commodity lands may do likewise (if game species populations are locally boosted). Virtually any pattern and intensity of forest management activity, including passive management, af-fects wildlife populations, invariably favoring some species and populations and impeding the success of others (Hunter 1990; Diamond 1992; Alverson and Waller 1993). Benefits to wildlife can be cited as proof in support of nearly any plan.

The abstract concept of multiple use can be fully satisfied in an almost in-finite variety of forest settings and uses—it is virtually impossible to manage lands without serving a multitude of uses. The real substance, then, of any particular multiple-use management regime is the relative dominance of var-ious uses. To assess the real consequences of a multiple-use plan for a forest, the question is not whether particular lands are or are not included in the multiple-use plan, but what are the quantities of each use and the extent to which each use dominates the landscape over time and space.

Clawson came to similar conclusions about the relationship of dominant and multiple uses:

> If dominant use is interpreted to mean that some particular use of the forest completely dominates all other uses to the extent that other uses are ignored or neglected or suppressed, then most of us would reject dominant use as a forest land use principle. But if it means that one use of the forest is primary in the sense that this use provides the major purpose of the planning and management of the forest, with other uses considered and adjusted to it as far as reason-ably possible, then dominant use is a sound method for objective of forest management . . . . On this basis, multiple and dominant use are not exclusive forms of forest management, but dominant use is a particular kind of multiple use. (Clawson 1975, p.41)

If we accept the notion that multiple use is a question of relative domi-nance, with some uses exerting very little dominance (and hence allowing many other compatible uses in the same or nearby time periods and/or areas), while other uses are much more dominant (precluding more uses for greater periods of time and space), then we must examine the relative dominance of uses in the historical practice of multiple use by the Forest Service and how this concept might be viewed in light of the substantial ecological incompatibilities (i.e., the domination of ecological processes and composition by the forces resulting from landscape disturbance) that we have identified in Part II.

The Forest Service Doctrine of Multiple Use

An important aspect of multiple use is the way in which it is conceptualized in the statutes and considered in the abstract. However, multiple use is also a doctrine. If one hears the term used in a speech by a Forest Service line officer or reads it in forest planning documents, it is this additional gloss of meaning above and beyond the abstract idea of mixing uses that is important to an understanding of the term.

The doctrine of multiple use was first formally considered by the Society of American Foresters during World War II (Clary 1986). Its main architect, however, has been the Forest Service, which first sought to use the concept as a rebuttal to increasing criticism during the late 1950s for its expanding timber program. Given this usage in the postwar years, it is understandable that some within the Agency are reluctant to acknowledge that any uses may be dominant within the multiple-use framework. Hence, it is, at its origins, a statement by the Agency that uses other than timber are going to be given considerable weight.

Because of the fact that nearly any plan, no matter how heavily timber may be dominant, has other benefits, it was and is easy to say that a timber sale or forest plan complies with multiple use. That heavily timber-dominant plans could justifiably be said to be within multiple use was a virtue of this very fluid concept. The high degree of compatibility between timber and game management has strongly reinforced the argument that multiple use can be practiced without overwhelming dominance of one purpose, and has encouraged the myth that active timber management is beneficial to wildlife (Chapter 8) and, by extension, beneficial to all aspects of a vigorously growing and "healthy" forest condition (Chapter 12).

Forest Service Multiple Use is achieved through a series of local dominant-use projects

Statements by Forest Service staff make it clear that they do not expect every local area of forest land to provide simultaneously for all valid forest uses (Thornton 1974; Krugman 1990). If one defines dominant uses as those that exclude other uses from occurring simultaneously in the same local area, then Forest Service Multiple Use at a forestwide (or regional) level is a locally dominant use system. Most local projects of the Forest Service also meet a second, slightly different definition of dominant use:

> The dominant-use concept implies several uses, all of which may (and generally do) receive some active management attention, but with one use (or in the case of wilderness, one might say "nonuse") predominating in terms of management policy. (Robinson 1975, p. 56)

For example, within a project designed to extract timber from a local area that also provides other valid uses, such as game, timber production can be called a dominant use, since it drives planning decisions at the local level. Likewise, management for wilderness in federally designated Wilderness Areas constitutes a dominant use since concern for other uses must be subordinated to this goal during planning efforts. Such a system of locally dominant projects provides for a balance of forest uses if each valid forest use is sufficiently addressed somewhere in the forest area being managed. The problem with Forest Service Multiple Use, relative to biodiversity concerns, is that timber (or less often the production of game, livestock, or recreation) drives the planning process within the majority of local projects so that biodiversity is addressed only after fundamental decisions about the extent, pattern, and intensity of landscape disturbance have been made.

Sunken Camp: An example of timber dominance

During the fifth year of our involvement with the National Forest planning effort in Wisconsin, we were asked by the Chequamegon National Forest to respond to a planning document for a 9800-hectare project named Sunken Camp. The Sunken Camp project presented unique opportunities for the Agency and was driven by a desire to harvest large, mature stands of jack pine before they aged beyond the point of having pecuniary value (Kick et al. 1991). In order to harvest these stands in an economically efficient manner, the Agency preferred to concentrate cutting into three blocks, each 324–372 hectares in size, much larger than permitted by their standard 40-acre (16.2-hectare) limit to clearcut size (Kick et al. 1991). To carry out this plan, the Agency made overtures to Wisconsin environmental groups to ensure that none would contest the proposed exemption to the 40-acre clearcut rule.

We did not oppose this exemption, and in fact welcomed it as a forest planning opportunity in which a substantial portion of the Chequamegon National Forest, the 9800-hectare Sunken Camp Opportunity Area, might be zoned for relatively intensive commodity production, in theory allowing other substantial portions of the Forest to be exempted from logging pressure. The outcome was otherwise. The Agency explicitly refused to make any commitment to undisturbed forest blocks of equivalent size to the Sunken Camp project either within the project or elsewhere in the Forest. Any possibility of old-growth designation was deferred to future analyses (Kick et al. 1991).

Alternative 2, the Agency's preferred alternative for the project, was summarized as placing "an emphasis on providing large units of habitat for plant and animal species" (Kick et al. 1991, p. [i]). This, and many other statements throughout the project's environmental impact documents, might lead one to think that concern for wildlife and biodiversity influenced management decisions in an important way. However, the project did not proceed by first deciding which species were underrepresented elements of biodiversity within the forest or the region, nor did it proceed by considering which native species could be produced by returning some of Sunken Camp Opportunity Area to its original pine barrens vegetation, and then deciding on a course of management to produce this coterie of species.

Just the opposite was true. Management was determined on the basis of timber needs, after which lists were compiled of species benefiting from that chosen management regime. Put another way, the welfare of nongame wildlife had no bearing on the decision-making process but was only considered after the fundamental management decisions had been made. The species benefiting from the young jack pine habitats created by the chosen alternative are common in the Chequamegon National Forest and the region as a whole. Populations of disturbance-loving species like deer, whose populations were already over management goals at the time these management decisions were being made, were predicted to increase under the chosen alternative. Yet, the Sunken Camp project rejected the opportunity to provide long-term habitat for the sharp-tailed grouse, a native open-lands (barrens) species whose habitat was usurped by the planting of jack and red pines, the harvest of which now drives this project (Kick et al. 1991, p. 137).

The example provided by the Sunken Camp project, perhaps better than any other we might offer, demonstrates timber dominance in the decision-making process under Forest Service Multiple Use. Despite the appealing claim that the project will place "an emphasis on providing large units of habitat for plant and animal species," such habitat will be provided only if it fits into a regime of cyclic habitat disturbances determined by timber objectives. Biodiversity did not drive planning decisions in the Sunken Camp project, yet the wildlife favored by planned activities were used to justify the project as multiple use and to showcase it as a New Perspectives project (below).

Forest Service Multiple Use assumes dominant uses must
be commodity uses

The degree to which Agency culture has shaped its concept of multiple use
has been commented on by numerous authors (Barney 1974; Clawson 1975;
Clary 1986; G. Robinson 1988). Valid use, under Forest Service Multiple
Use, has come to mean commodity production, usually timber production in
particular. Areas subject to passive management, or within which no timber
harvest is allowed, are considered to be underutilized, wasted lands.
Wilderness, either for recreation or for biodiversity, is not viewed as a valid
dominant use, and such lands are considered as apart from multiple-use (i.e.,
logged) lands (see further discussion in Chapter 14). Similarly, management
for biodiversity is not considered a suitable dominant use under Forest
Service Multiple Use.

The Forest Service's increasingly timber-biased approach to management
emerges conspicuously in historical and current data regarding how it allo-
cates its resources (Steen 1976; O'Toole 1988; Sample 1990; Hirt 1991).
Despite adhering publicly to the concept of a balance among multiple uses,
the Agency has, in fact, increasingly embraced commodity production as its
primary function. This began in earnest as demand for housing and other
forest products soared after World War II (Fig. 8-1). The increase in timber
harvests and clearcutting clearly represented a movement toward timber as a
dominant use across National Forest lands (Shepherd 1975; Clary 1986; G.
Robinson 1988).

A strong timber emphasis is also apparent when we examine funding and
personnel levels within the Forest Service. The distribution of Congressional
appropriations among various Agency programs illustrates the dispropor-
tionate emphasis roads and timber harvests have received over three recent
decades (Fig. 9-1). This bias also is evident in the distribution of personnel
within National Forests. For example, within the Wisconsin National Forests
in January 1990, 33.1% of the forest staff worked in forestry (primarily silviculture and timber sales) and 15.9% in engineering (primarily road construction). In contrast, only 4.3% of forest personnel worked in any aspect of
biology, and the majority of these were working in wildlife (primarily fish and
game management). While many more biologists and some ecologists are
employed by the Research branch of the Forest Service, this heavy bias in
staffing and expertise within individual forests makes it difficult to inject concern for ecological processes or the habitat needs of nongame species into
routine forest planning and management.

Overall land use allocations within the National Forests also clearly reflect
the domination of timber over other uses. Within the Wisconsin National
Forests, the long-term management plans of 1986 assigned an overwhelming

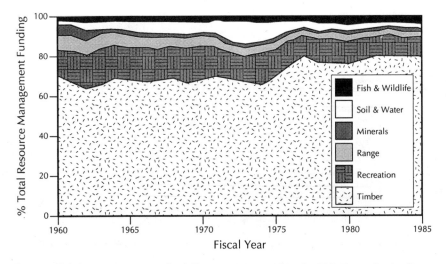

Fig. 9-1. Relative appropriations for different activities within the U.S. Forest Service from 1960 to 1985. Note high and increasing fraction of expenditures for activities related to timber harvesting. [Figure redrawn from Sample (1990).]

majority of the total land base to clear- or selective-cutting treatments (Fig. 9-2).

Less than 2% of the Chequamegon National Forest's area currently falls within Management Areas free from timber harvest (USDA Forest Service 1986a). Of the approximately 98% of the forest within timber management areas, not all areas will be directly affected by timber harvest during the life of the management plan. In fact, only 54% of these nonreserved lands are currently in the "timber base." One cannot conclude, however, that almost half of the forest's area is free from the effects of timber harvest. The areas not included in the timber base are small and scattered. Furthermore, many of these areas will be included in the timber base in future years as their trees age and other physical and economic factors change. From a conservation point of view, these areas are extremely fragmented, will not be allowed to proceed to old-growth conditions, and are highly subject to pernicious area, edge, and isolation effects (Chapters 4–6). In this sense, they are no less part of the timber base than the spaces between the trees to be harvested.

While other national forests allocate far more to Wilderness Areas and other types of low-intensity management, these specific data and system-wide totals evidence a clear priority for intensive timber management. The default appears to be timber management, with other uses emerging strongly only in areas where timber uses are either proscribed by law (e.g., Wilderness Areas) or blatantly incompatible with aesthetic or recreational values.

To accommodate concerns for rare habitats and elements of diversity

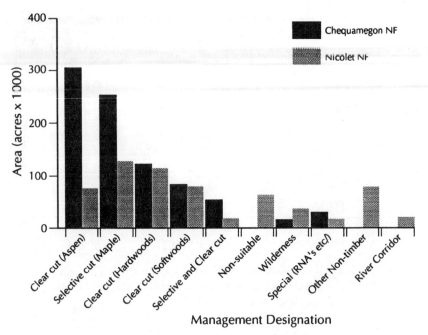

Fig. 9-2. Allocation of Wisconsin National Forest lands to various uses. Note high fractions of land area dedicated to primary uses involving timber harvests. [Data from long-term management plans for the Chequamegon and Nicolet National Forests (USFS 1986a,b).]

possibly threatened by timber cutting or other development, the Forest Service relies on both existing Wilderness Areas and Research Natural Areas (RNAs). Wilderness Areas, however, almost always represent areas chosen for their aesthetic or recreational value rather than their value as biotic habitat (Chapter 14). While RNAs are specifically designated to protect biotic values, they are generally few in number, isolated, and quite small—all features that limit their ability to function effectively to protect biodiversity. In surveying 213 RNAs nationally, Noss (1991a) notes that 93% are smaller than 1000 hectares. In Wisconsin, existing and proposed RNAs are even smaller, with a mean size of only 56 hectares.

New Perspectives as a public relations vehicle

How has the Forest Service responded to concerns over the dominance of timber over other valid multiple uses? Sadly, leadership within the Forest Service has adopted the official line that conventional forestry poses little or no threat to natural values and has tended to treat challenges to that position as attacks on its professional competence and discretion (Chapters 12 and 13). They have also tended to treat issues surrounding the management of biotic resources as an image problem rather than a legitimate and important debate over how scientific principles should be applied. Thus far, we have seen the regular development of new elaborate and assertive public relations campaigns and a handful of conspicuous demonstration projects but few on-the-ground changes in Forest Service management practices. Thus, one encounters a disturbing dissonance between public pronouncements and continuing efforts to sustain high cut volumes that many consider inconsistent with maintaining threatened or declining populations of wildlife.

Because of the intense criticism leveled at the Forest Service during the first round of forest planning in the mid 1980s, the Agency reformulated its multiple-use policies to better take environmental concerns into account. The resulting initiative, New Perspectives in Forestry, was developed in the upper echelons of the Forest Service, and as such, is a carefully constructed chimera of politics and science. Hal Salwasser, a wildlife biologist, was one of its chief architects and became the main spokesperson for New Perspectives (Salwasser 1991a,b; 1992).

It is difficult to argue with the vague and lofty goals of New Perspectives, yet these goals have only poorly translated into tangible changes in the day-to-day management of National Forest lands. New Perspectives literature begins with broad discussions of world population pressure, conversion and degradation of forests worldwide, and the profligate use of forest products in the United States. As citizens, New Perspectives challenges us to take responsibility for our consumptive behavior and not pass environmental costs on to faraway countries whose forests are plundered for our use. Instead, we are told to produce our own forest products domestically and in an environmentally responsible manner that respects all values forthcoming from forest ecosystems.

There has been much fanfare over this initiative. We have been put on notice that this represents a radical change in Agency thinking, ushering in a new and enlightened age of resource management, but many remain skeptical. In the Midwest, when press releases were distributed trumpeting the virtues of New Perspectives management, line officers had little idea what it was or how to define it (Evans 1990). Because the reality of New Perspectives is still taking shape, it can only be defined by the tangible projects and accomplishments it has spawned. Its goals are now exemplified by a myriad of projects at the local and regional levels, some of which are research projects and many of which were ongoing prior to the New Perspectives initiative.

As uncontrovertible as it might sound in theory, the New Perspectives initiative shows few signs of changing the realities of National Forest management. While future decision-making processes will appear different, the great influence of commercial timber production on planning has not been significantly altered (Frissell et al. 1992; Lawrence and Murphy 1992; Chequamegon National Forest staff, pers. comm.) Two examples follow, the Shasta Costa and the Sunken Camp New Perspectives projects.

The flagship project of the New Perspectives initiative was the Shasta Costa Integrated Resources Project on the Siskiyou National Forest in Oregon (Camarena et al. 1990). Announced by a full color, glossy booklet, it was to be the paragon "to illustrate how New Perspectives can work on the ground" (Camarena et al. 1990, p. 1). In short, the booklet described the new and clever ways in which commodity production (i.e., logging) would be blended with concern for environmental protection to produce a middle path between the allegedly unreasonable "preservation-first" or "products-first" approaches. To do this, many of Franklin's New Forestry technologies (Chapter 10) would be employed to ameliorate damage on harvested lands.

The reality of the Shasta Costa project took a different course, as described by forest activists involved in the process (Britell 1992; Frissell et al. 1992; Zuckerman 1992). After many months of cooperative work with forest activists, the Agency chose the ecologically benign Alternative 3 of the Draft Environmental Impact Statement, a "radical departure from traditional approaches" that "would provide for entering the [Shasta Costa] basin without roading its roadless area" (Britell 1992, p. 9). A timber volume of about 10–11 million board feet would be harvested, slightly less than that scheduled under the existing Forest Plan (Camarena et al. 1990; Britell 1992). As in our Wisconsin experience, the Draft Plan then disappeared from sight for a number of months and reemerged in a radically different form from that of prior agreements and expectations. In the Shasta Costa, the amount of old growth to be cut increased by 600%, the core of the roadless basin was to be roaded, and the timber volume increased to 14–15 million board feet. Furthermore, forest activists gradually became aware that the Shasta Costa project agreements had in no way altered the timber quotas set by the Forest Management Plan. Any shortfall from the Plan's target of 32 million board feet would be made up in the Shasta Costa basin after the New Perspectives project was over, but within the 10- to 15-year life of the Plan.

Somewhat ironically, the Sunken Camp project of the Chequamegon National Forest (discussed above) has also been touted as a New Perspectives project whose aggregation of clearcuts "reduces fragmentation and allows other large areas to be managed for continuous forest cover to provide habitat for certain species that require interior forest conditions" (USDA Forest Service, Eastern Region 1992a, p. 22). But, the "large areas" (each 240–325 hectares in size) within the Sunken Camp project were not designated to become mature forest or old growth, as a casual reading of this statement might suggest. Nor were any other areas within the Chequamegon National Forest in any way designated, specified, or agreed upon to be managed for old-growth conditions. The "large areas" within Sunken Camp were simply areas that had "continuous forest cover" and would not be cut in the next 10 years. In fact, these include extensive acreage in young aspen and pole-sized northern hardwoods scheduled to be cut within the next 11–50 years under the guidelines of the Chequamegon Forest Plan.

Other New Perspectives projects in the Eastern Region (Forest Service Region 9) include visual quality enhancement by thinning cuts and white pine underplanting, the use of grazing by cows and sheep to maintain the artificial open land ecosystem at a Confederate Civil War site, the restoration of waterfowl habitat, and an interpretive trail through an old-growth pine stand (USDA Forest Service, Eastern Region 1992a,b). As for projects which combine timber production and preservation of old-growth characteristics on the same acreage, there are but few, including the Hayes-Bay timber sale on the Shawnee National Forest in which logging mimics natural gap-phase dynamics, and an attempt by the Ottawa National Forest to increase the amount of coarse woody debris and dead and dying trees on some harvested stands (USDA Forest Service, Eastern Region 1992b).

For eight years, we have raised issues of edge effects due to relatively high population densities of white-tailed deer on National Forest lands in the Midwest, an ecological factor central to landscape-level planning for old growth (WFCTF 1986; Alverson et al. 1988), the welfare of certain rare species (such as orchids—Miller et al. 1992), and the distribution and frequency of timber harvest (shelterwood and selection cuts included). We have also called for extensive sampling efforts directed toward poorly known organisms, such as lichens and invertebrates, in mature and old-growth stands on National Forest land (WFCTF 1986; Solheim et al. 1991), so as to make better-informed management decisions relative to relaxation times and edge effects. Yet in the listing of New Perspectives experimental and demonstration projects (USDA Forest Service, Eastern Region 1992a,b), not a single project was cited whose purpose was to gather descriptive or experimental data to respond to these concerns.

The differences between the rosy impression given by New Perspectives literature and the management decisions embodied by Forest Plans underscore the public relations nature of the New Perspectives program. New Perspectives is not a set of guidelines for scientific forest management but rather a means of navigating existing Forest Plans through politically turbulent waters. Each Forest Plan is a legal document with specific goals (timber targets, maps of land-use zoning, etc.). In contrast, New Perspectives literature, projects, and commitments are "evolving" (Camarena et al. 1990; USDA Forest Service, Eastern Region 1992a,b), with no enforceable responsibility by the Agency as to their scientific veracity or institutional commitment.

New Perspectives and science

The scientific component of the New Perspectives initiative is relatively small. Furthermore, as was noted by Noss (1991a), most New Perspectives projects are ongoing experiments only. They as yet offer few data subjected to peer review in the scientific community at large to confirm the idea that New Perspectives offers a viable management model to meet the needs of commercial forestry and protect regionally significant elements of biodiversity at the same time. New Perspectives is superficially similar to Franklin's New Forestry (Chapter 10), but it is primarily a public relations effort with little scientific integrity to date.

The foremost of the New Perspectives assumptions as yet to be supported by critical scientific review is that land managers can address all valid biodiversity concerns on lands managed for timber. If this assumption were true, we could solve the biodiversity problem simply by carrying out more sensitive forestry practices within the context of traditional Forest Service Multiple Use without challenging the timber dominance inherent in this management philosophy. As stated by Noss (1991a), we consider this central assumption of New Perspectives to be scientifically unfounded for two basic reasons: (1) We simply do not know enough about nature to actively and responsibly manage for all important aspects of biodiversity (Chapters 2–7), and (2) The little that we do know strongly suggests that not all biodiversity concerns can be addressed on lands managed for timber dominance or in the relatively small acreage of lands reserved from timber harvest. Thus, New Perspectives has concluded prematurely, and without scientific justification, that new biological reserves—forest zones from which logging and other significant, artificial disturbances are excluded—are not needed to conserve biodiversity.

It is one thing to argue that land allocation for such new biological reserves will be difficult for political reasons (Harris 1984; Thomas and Salwasser 1989) but it is something very different to argue that reserves are unnecessary from a scientific viewpoint. Land managers need these biological reserves to assess the success of New Perspectives experiments at the landscape level and because the results of New Perspectives experiments may not support the Agency's already-drawn conclusions:

> Scientists shudder to think of experiments without controls, but this is precisely what happens with most of our land management. Because many effects of land management are expressed at the scale of the landscape, we need control areas large enough to encompass

landscape-level processes such as disturbance propagation, patch dynamics, and between-community fluxes of organisms and materials. (Noss 1991a, p. 120)

And,

Salwasser misrepresents the views of Noss (1991a). The argument for the role of wilderness in conservation does not derive simply from the assumption that "nature-knows-best." It springs from the assumption that despite good intentions (and a few bad ones), humans, and bureaucratic systems like the Forest Service, do not always know what is best. We need representative natural systems that are not deliberately manipulated for what may later turn out to be naive purposes of resource "improvement," both in case our assumptions are wrong, and in order to maintain a benchmark against which the success of manipulative management experiments can be judged. (Frissell et al. 1992, p. 462, footnotes omitted)

Ecosystem Management: New Perspectives with a new paint job?

Agency staff are quick to point out that the New Perspectives initiative is changing. The term Ecosystem Management has now been substituted for New Perspectives (K. Holtje, pers. comm.). New projects continue to arise under the auspices of this most recent version of Forest Service Multiple Use, such as the Ecological Classification System (ECS) program (Hoppe 1990; Chapter 7). Such projects are to be lauded as research efforts. However, available data and data proposed to be collected by the Agency in Wisconsin, whether put into an ECS computer database or not, are limited in scope. The data do not address many of the ecological factors necessary to confirm that we can actively manipulate most of the landscape while maintaining all important elements of biodiversity, as New Perspectives and Ecosystem Management suggest. Nor is it clear that these new scientific data will have a significant impact on actual management decisions within the context of Forest Service Multiple Use. Despite the fanfare of New Perspectives and Ecosystem Management, the reality of land management under Forest Service Multiple Use has changed little.

One final item should be mentioned for eastern conservationists considering New Perspectives and Ecosystem Management. The fine print of New Perspectives literature specifies that if timber targets are not met because of reduced harvest due to the mitigation efforts, more land will have to be brought under active timber management (Camarena et al. 1990). Thus, not only will there be no expansion of forest acreage reserved from commercial timber harvest, there will likely be many cases in which the area to be harvested will be *increased.* We'll have light-touch forestry, but it will be, on average, over a wider area than that directly affected by timber harvesting today

Summary

The most recent formulations of Forest Service Multiple Use, New Perspectives and Ecosystem Management, represent an adaptation to changing scientific and political views. For cultural and historical reasons, however, these hybrid management–public relations initiatives are predicated on the dominance of timber and other commodity products in planning efforts and therefore promote forest management tied to the active manipulation of vegetation.

Years ago, many forest managers made the erroneous assumption that young forest habitats were good for all wildlife (Chapters 5 and 8). New Perspectives and Ecosystem Management put a new face on this traditional assumption of Forest Service Multiple Use: We can make lands managed for timber dominance good for all wildlife if only we apply new technologies to protect what would otherwise be harmed.

Scientific confirmation of this assumption will take decades. Research efforts carried out under the New Perspectives and Ecosystem Management initiatives represent a step toward concern for biodiversity. However, the degree to which New Perspectives and Ecosystem Management allow timber dominance to masquerade as biodiversity protection is disturbing. These initiatives are used to justify dominance by timber in the majority of the Agency's local projects by proclaiming that all reasonable concerns over biodiversity have been (or will be) addressed.

Fortunately, other forest management models have been developed outside of the political and intellectual constraints of Agency culture. These recent models, representing a broader range of possibilities for both diversity protection and commodity production, are reviewed in the next chapter.

10

What's New in Forest Management?

Most conservation and commodity production groups consider the whole range of multiple uses as defined in the Multiple-Use Sustained-Yield Act (MUSY) to be valid uses of National Forest land, as do we. Taken together with the general agreement between land managers (in practice, if not in word) that a locally dominant-use, forestwide multiple-use management model is acceptable, it is perhaps difficult to understand all of the recent commotion over biodiversity concerns. At issue is the scale and pattern at which the various multiple uses are provided for in forest management.

Over the last few decades, there has been a growing awareness in the scientific and conservation communities that the diversity of native species and indigenous ecological processes is not adequately protected by the existing management regimes on many public and private forest lands (Chapters 4–9). This presents a dilemma for multiple-use management as it has been practiced for the last 50 years, a painful realization that there are new factors that must be accounted for. In response, a number of new multiple-use management models have been proposed that address biodiversity concerns while simultaneously attempting to minimize the disruption of human economic and social activities.

Given the incredible latitude of interpretation of multiple use under MUSY, it is not surprising that there are some profound differences among these new multiple-use management models. Some focus on maximizing biodiversity protection while creating as little disruption of human communities as possible; others the reverse. Some explicitly acknowledge the fact that many uses are locally dominant uses; others do not.

In this chapter, we provide brief descriptions of the multiple-use management models now competing to become the management paradigm for approximately 77 million hectares of National Forest lands and, by example, millions more.

The Excellent Forestry Model

Gordon Robinson's recent book, *The Forest and the Trees: A Guide to Excellent Forestry*, describes a forest management model advocating selective (individual tree) cutting and uneven-aged management to minimize ecological damage to lands subject to commercial timber harvest (G. Robinson 1988). Robinson opposes the use of clearcutting and other forms of even-aged management, such as seed-tree and shelterwood cutting, favoring instead other, less intensive forestry techniques to reduce forest disturbance and better protect soils and watersheds.

There is little question that Robinson's proposed silvicultural methods would reduce soil compaction, soil erosion, and associated biological changes wrought upon a harvested stand. However, the Excellent Forestry Model suffers from two significant flaws: Both the criteria used to judge "excellence" and the scale at which the model is formulated lack biological sophistication.

This model assumes that its proposed silvicultural techniques alone suffice to protect the full range of biodiversity values, species, and processes in stands subject to logging. Readers of Robinson's text may conclude that we possess adequate knowledge and appropriate management technologies to assume responsibility for all biodiversity concerns on commodity-production acreage. However, this is far from certain, as discussed here (Chapters 2–6) and elsewhere (Leopold 1941; Noss 1991a; Willers 1992). Robinson's book conceptualizes certain aspects of forest ecology from a narrow silvicultural point of view. Although the sustainable forest maxim is "Don't remove anything that is not replaceable within a rotation," there is little indication that specified criteria such as "older timber; multistory, multispecies stands; protection of soil and watersheds; and reasonably neat and orderly logging" are sophisticated enough to ensure that these silvicultural methods will sustain all elements of biodiversity on a local basis.

The second flaw concerns theoretical and practical aspects of scale. The Excellent Forestry Model addresses only site-specific mitigation of negative biological effects caused by timber harvests, not pervasive or cumulative landscape impacts. Thus, from a landscape perspective, this management strategy represents a suboptimal solution to the retention of biodiversity. The Excellent Forestry Model could doubtlessly be strategically applied to better preserve biodiversity at local scales, but it should not be adopted at a landscape scale unless timber production targets are significantly lowered. Without an overall reduction in forest output, the application of this model uniformly across a forest landscape would force more land into the active timber base. In other words, the fault with the Excellent Forestry Model relative to biodiversity concerns is that it tries to accommodate all valid forest uses at the local level rather than accept a locally dominant-use, regionally multiple-use paradigm (see review by Krugman 1990).

In summary, Robinson's Excellent Forestry Model is well-intentioned and capable of reducing the local severity of timber harvests, but it should not be embraced as a general model for management of eastern forest landscapes, especially if timber activities are forced to expand into reserved lands.

The New Forestry Model

This model grew out of years of silvicultural and ecological research in old-growth stands in western forests, especially in the Andrews Experimental Forest near Blue River, Oregon. Many scientists participated in this National Science Foundation–funded effort initiated in 1970, including Jerry F. Franklin, who has come to be a major spokesperson. Years of study of old-growth forest ecosystems led Franklin and colleagues to develop a forest management model aimed at protecting biological diversity within a landscape managed for commercial timber production. Like the Excellent Forestry Model, New Forestry adopts management techniques to mitigate damage to both the physical and the biological environments. To protect the "biological legacy" of harvested sites, efforts are made to increase the probabilities that desirable plants and animals, particularly those of later successional stages, remain on site during and after harvest or can recolonize from surrounding habitats. For instance, patches of uncut trees ("green tree retention") and coarse woody debris (downed logs and branches) are left to provide continuity of conditions and habitat for salamanders, insects, and fungi (Franklin 1989, 1990; Gillis 1990). Finally, and importantly, the arrangement of timber cuts is also considered at a landscape level to ensure that some old-growth reserves persist at all times in the landscape.

The New Forestry Model is based on extensive biological knowledge of western forests and is faithful to scientific principles. Seemingly arcane but critically important factors such as the ecological guild structure and dispersal capabilities of canopy invertebrates are considered, as are complicated ecological interactions between mycorrhizal fungi, mammals, and woody plant regeneration. Because of this biological sophistication, the model is more conservative in its promises of what it can and cannot do than are its politicized derivatives, New Perspectives and Ecosystem Management.

This model explicitly notes the need for areas reserved from commercial timber harvest: "New Forestry does not eliminate the need for reserved areas, as has occasionally been suggested" (Franklin 1990). Franklin chastises environmentalists who rely only on "set-asides" (lands on which commercial forestry is barred) to protect biological diversity, yet he does not contend that all biological diversity concerns can be met in the context of the commodity-managed areas, even with the application of New Forestry techniques: "I'm not proposing that New Forestry be used as a substitute for preservation, but I think wherever we make a decision to cut timber, whether it's old-growth or other forests, we should consider New Forestry" (Franklin, in Gillis 1990, p. 560).

Numerous aspects of the New Forestry Model apply to managing eastern forests. We critically need extensive biotic inventories, especially of commonly overlooked organisms such as fungi and invertebrates within our remaining old-growth habitats (Chapter 7). With these inventory data in hand, eastern biologists and conservationists could draw on the ecological databases now established by New Forestry programs for species and genera of organisms which co-occur in western and eastern forests (e.g., old-growth lichens such as *Lobaria pulmonaria*—Thomson 1990; Lesica et al. 1991). Inventory and ecological data like these collected during the next few decades will allow us to predict with more accuracy the biological effects of active forms of forest management. In the meantime, New Forestry techniques such as green tree retention and increasing available coarse woody debris can ameliorate some of the known or suspected deleterious biological effects of logging.

In summary, the New Forestry Model proposes scientifically based, *experimental* techniques to improve biodiversity protection through mitigation of ecological damage on lands harvested for timber or fiber. Because of this focus, other issues related to forest biodiversity are not emphasized. In particular, the degree to which extant or potential biological reserves offer sufficient protection to resident biota and ecological processes are not considered in depth. In contrast, the next two models presented here begin not by considering how biodiversity mitigation on lands slated for timber production should proceed, but by asking instead which lands might be zoned for timber dominance and which for other dominant uses (including biodiversity) at a landscape level.

The Multiple-Use Module (MUM) Model

After an extensive review of the biology of the Willamette National Forest, and of island biogeography theory, University of Florida biologist Larry D. Harris proposed a landscape-level system of long-rotation habitat islands, each with an old-growth core reserved from commercial timber harvest (Harris et al. 1982; Harris 1984). His management system is explicitly concerned with old growth as a component of overall forest diversity:

To the extent that remaining old-growth Douglas fir ecosystems possess unique structural and functional characteristics distinct from surrounding managed forests, the analogy between forest habitat islands and oceanic islands applies. Forest planning decision variables such as total acreage to be maintained, patch size frequency distribution, spatial distribution of patches, specific locations, and protective measures all need to be addressed. (Harris 1984, p. 6)

It must also be taken into account that the thrust of this work is not directed toward a system of nature preserves, but rather toward a forest management strategy that depends heavily upon, and in turn conserves, the unique characteristics of old-growth ecosystems (Harris 1984, p.8).

Given the impending conversion of many western, old-growth forest lands to younger, more intensively managed stands, Harris's model seeks to maximize the biologically effective size of the islands of old-growth habitat that will remain after the old-growth forest matrix around them is removed. This is to be done both by maintaining old-growth habitat islands of at least a minimum size and by minimizing the habitat differences between each old-growth island and the surrounding forest habitats. Concentric rings around each old-growth island would be zoned into areas of reduced human-caused disturbance, with increasingly severe disturbances allowed as one moved away from the center. The innermost ring would consist of pie-shaped wedges (Fig. 10-1), each of which would be harvested at a frequency one-third that of the normal short-rotation cycle (e.g., at 320-year intervals rather than 80). Outer concentric rings around this innermost, long-rotation ring would be zoned for increasingly intensive commodity production, in transition to the surrounding human-dominated landscape. This model can be applied in many non–forest conservation situations as well (Harris 1984; Noss 1987b). The general idea is one of zoning, in which core biological reserves are surrounded by concentric zones with a centrifugal intensification of human disturbances (cf. Noss 1993).

Other aspects of the MUM Model were developed specifically for the protection of western old growth and cannot be directly incorporated by planning models for eastern forests. For instance, Harris's model deals only indirectly with old-growth restoration because the context of Harris's work was a landscape in which the forest matrix still contained a substantial amount of old growth. Harris was looking at the pre-isolation phase of these old-growth islands and hoped to influence the pattern into which relatively large forest stands would (inevitably) be cut, so as to retain as much ecological integrity

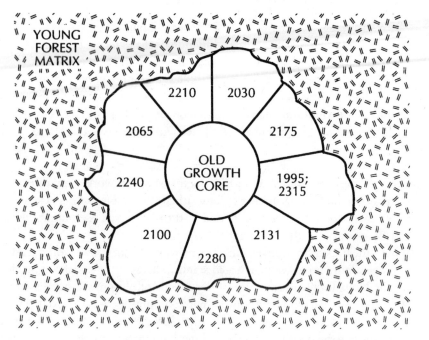

Dates of timber harvest for 320-year rotation sequence

Fig. 10-1. The Multiple-Use Module Model of Harris, who proposed that a central "core" of reserved old growth would be surrounded by a ring from which pie-shaped blocks would be periodically clearcut. Here, the first block would be cut in the year 1995, the next in 2030, then 2065, etc., until the year 2315, when the first block would be revisited and recut. The very long-rotation cutting schedule (in this case, 320 years) and pattern of cutting in large, contiguous blocks partially but continually surround the old-growth core with older forest stands, presumably buffering it from the surrounding matrix of disturbed young forest. [Redrawn from Harris (1984) with permission from the University of Chicago Press.]

as possible. A very different situation occurs in most eastern forests, where the presettlement forest fabric has already been rent and removed, leaving, in most cases, only small fragments. Whereas Harris considered the potential effects of island biogeography relaxation times on the biota of old-growth habitat islands to be isolated in the future, eastern conservationists must address relaxation times that have been ongoing for at least a century. What we must ask is which old-growth components have survived in our habitat islands, how unique are they in terms of regional diversity, and how long can they be expected to persist if these habitat islands remain isolated from other old-growth stands?

Harris suggested a log-normal size class distribution of old-growth islands, arguing that such distributions are commonplace in nature (cf. Hunter

1990). This size class distribution implies that the old-growth archipelago should consist of many small and medium-sized habitat islands (of 20 to 300 hectares each) and fewer large islands, including single islands of up to 12,500 hectares (Harris 1984). Harris did not consider old-growth islands greater than this size for both political and biological reasons:

> If these [northwestern] old growth ecosystems are expected to func-
> tion independently of their surroundings, then their size might
> need to be so large that their conservation would be outside the
> realm of existing conditions or current political or economic re-
> straints. (Harris 1984, p. 128)

In reality, no old-growth forest stand or ecosystem operates independently of its surroundings, though smaller stands will on average be more greatly influenced by biological factors outside of their boundaries (Chapter 5). In the MUM Model, Harris makes the reasonable assumption that surrounding core old-growth areas with buffer rings subject to long-rotation timber harvest will increase the biologically effective size of each old-growth habitat island. However, while these long-rotation, periodically logged buffer areas undoubtedly help maintain some old-growth species and processes, it is not yet safe to conclude that they can serve as a direct substitute for forest habitats undisturbed by commercial logging (Duffy and Meier 1992; Duffy 1993a,b; Elliott and Loftis 1993; Johnson et al. 1993; Bratton 1994; Matlack 1994).

In the eastern U.S., most remnant old-growth stands are very small (20 hectares or less—much smaller than those considered by Harris; see Fig. 1-2) yet will often retain some species and processes characteristic of old growth, despite the time elapsed since their isolation from the original old forest matrix. Advocates of biodiversity reserves in eastern forests should consider increasing both actual and effective sizes of selected, remnant old-growth habitat islands through active or passive restoration efforts, as discussed in the next chapter. Although the opportunities to expand old-growth habitat islands are subject to complex political and social considerations, eastern conservationists should begin by asking the important scientific questions: how big, what shape, what distribution, and what management should ecologically effective biodiversity reserves have?

In summary, the MUM Model proposes a landscape-level zoning of the landscape into areas of greater and lesser silvicultural intensity. The conservation of old-growth forest stands and natural ecological conditions would be the primary management goal (or dominant use) within core zones, while commodity production would be addressed in surrounding forest zones. The most intensive landscape disturbances would be separated from the core old-growth reserves by concentric areas with intermediate disturbance regimes.

The Dominant Use Zoning (DUZ) Model

In 1986, the Wisconsin Forest Conservation Task Force, a group of biologists, economists, environmentalists, lawyers, and planners, formulated a generalized model for landscape-level planning efforts in eastern forests (WFCTF 1986). Here, we briefly describe the general outlines of the model. A more extensive description and justification of the biological reserves called for by the model appears in Chapter 11.

The DUZ Model was developed with regard to the protection and restoration of interior forest conditions in the upper Midwest. It is a locally dominant use, regionally multiple-use model focused on the protection of biodiversity through the creation of new, large biological reserves (Diversity Maintenance Areas, or DMAs) in which all forest uses compatible with the long-term maintenance of interior forest conditions would be allowed. In contrast, commercial timber harvest, new road construction, and new wildlife openings would be specifically excluded from the 15–25% of overall forest area designated as DMAs because all of these activities produce significant amounts of young, disturbed forest habitat.

This model is most similar to Harris's MUM Model in that its focus is the protection of old-growth forest reserves rather than mitigation of adverse impacts to biodiversity on lands directly subject to logging. It differs in that the DMAs would be largely restored to mature and old-growth conditions rather than already existing in that state, as is the case in some western forests. The proposed DMAs would also be significantly larger than the old-growth islands envisioned by Harris.

The Dominant Use Zoning Model was developed with the assumption that both logging (including clearcutting) and the designation of new areas reserved from logging represent valid multiple-use management of eastern National Forests and other public lands. In view of this, technologies proposed by the New Forestry and Excellent Forestry Models to mitigate negative effects of timber harvesting are fully compatible with the Dominant Use Zoning Model.

Summary

Each of the multiple-use management models offers different assumptions and concepts potentially useful for the protection of biodiversity in eastern forests. Some models exhibit a great degree of compatibility with others, offering similar management recommendations but with a different emphasis.

The Excellent Forestry, Multiple-Use Module, and New Forestry Models emphasize methods of mitigation of adverse impacts to biodiversity on logged lands. The Multiple-Use Module and New Forestry Models also emphasize the need for areas reserved from logging and address the size, management, and distribution of these reserves as integral parts of multiple-use management. Similarly, the Dominant Use Zoning Model emphasizes the need for unlogged lands, in large contiguous blocks, as part of an overall multiple-use strategy that emphasizes protection of biodiversity.

Somewhat paradoxically, the Excellent Forestry Model comes closest to matching the assumptions of the Forest Service's New Perspectives and Ecosystem Management "models," perhaps because Robinson's own background was in forestry. These approaches imply that we know, or will know in the near future, how to actively manipulate forest stands for timber production while sustaining all valid aspects of biodiversity on the same acreage. Because claims of sustainable harvesting in other temperate and tropical ecosystems have often been shown to be invalid or overly optimistic once they were seriously evaluated scientifically (Vasquez and Gentry 1989; Duffus 1993; Ludwig et al. 1993; Phillips 1993; see also Linden 1991; Stevens 1993b), we urge readers to carefully evaluate similar claims made in this present context.

Given the uncertainty that Excellent Forestry, New Perspectives, or Ecosystem Management will adequately protect all aspects of biodiversity on acreage subject to commercial logging, we conclude that a significant proportion of each forest landscape should be designated for management under a natural disturbance regime, excluding as many human-generated sources of disturbance as is possible in modern times. This conclusion, fully compatible with the Multiple-Use Module and New Forestry Models, led the Wisconsin Forest Conservation Task Force to develop the Dominant Use Zoning Model, whose emphasis is on the development of large biological reserves in forest landscapes.

Zoning for Diversity

Assuming that all uses cannot reasonably be expected to coexist on the same acres, we face the significant question of deciding at which scale competing uses should be segregated. At one extreme, managers might attempt to protect sensitive species, or provide for timber or recreation, by planning in relatively small units of one to a few hectares in size. At the opposite extreme, one might envision entire National Forests, or even all the National Forests in a region, being allocated to one dominant use or another. Forest managers must address this problem of scale and justify their decisions regarding the allocation of lands within planning units. As a corollary, we need to restructure forest management to foster planning at the broad scale necessary to conserve biodiversity.

In this chapter, we describe a management model that zones multiple-use forest lands into large areas, each containing compatible dominant and subsidiary uses.

Forest Zoning at a Large Scale

When managers of eastern forests have agreed to allocate lands for the specific purpose of maintaining biotic diversity, they have usually done so on a small scale in the form of Research Natural Areas (RNAs), buffer strips, and relatively small wilderness areas of 10,000 hectares or less. Of these, only the RNAs are chosen specifically to protect species or community diversity, and even these have rarely been subjected to the type of minimum dynamic area analysis promoted in Chapter 4. Furthermore, these reserved areas are rarely integrated into any systematic network of reserves where proximity, contiguity, or the presence of connecting corridors is deemed important (Chapter 6). The result for most of the eastern states has been both limited segregation of competing uses and, where it does occur, rather fine-scaled segregation that tends to fragment or maintain the current fragmentation of natural habitats. National forest plans, for example, frequently stipulate that each Opportunity Area (OA) meet specific targets for maintaining particular habitat types such as old growth, ensuring the existence of many small and scattered patches of old growth throughout the forests. Furthermore, without formal designation and protection, these patches of old growth are subject to harvest in the next few decades. While such planning may formally meet Agency standards and guidelines regarding the protection of old growth, they often do little to protect those species most sensitive to forest fragmentation (e.g., those dependent on metapopulation dynamics or retention of natural disturbance regimes). In fact, such planning often contributes to further habitat fragmentation and homogenization of the landscape.

An alternative strategy requires that planning and the segregation of some forest uses be carried out at a much larger scale. In the preceding chapter, we outlined the Dominant Use Zoning (DUZ) Model, developed by the Wisconsin Forest Conservation Task Force (WFCTF 1986), and presaged by economist Marion Clawson's work:

> Suppose, for instance, that an effort were made to produce a sub-stantial volume of wood annually . . . on a relatively small acreage of forest land by means of intensive forestry, and that the rest of the land were devoted to forest uses other than wood production. What kind of a model or "scenario" might one devise? One such scenario might be based on forest ownership class and wood productivity classification, and one might put some of the forest land into inten-sive forest, some into careful "natural" forestry, some into a deferred harvest classification, and some into more or less permanent reser-vation. (Clawson 1975, p. 105)

And,

> [T]he essential point is that the great wood-production possibilities of intensive forestry on the better forest sites opens up enormous opportunities for multiple use forestry on a national scale. That is, some land can be used for one purpose, other land can be used for other purposes, each with various subsidiary uses; the whole would be very much a multiple use forestry management, even though every forested acre does not have every forest use every year. (Clawson 1975, p. 107)

Under the DUZ Model, National Forests and other forest lands devoted to multiple-use management would be zoned at a coarse scale into areas of greater and lesser intensities of certain management activities such as timber and game production, recreation, and biodiversity maintenance. As in the New Perspectives and Ecosystem Management models, biodiversity would be a consideration on all acres of such a forest, but it would be the primary man-agement priority within specific, mapped zones, termed Diversity

Maintenance Areas (DMAs). Timber, game production, and recreation would constitute the primary management priorities within other specific zones, allowing multiple use at forestwide and regional scales.

DMAs could be used both to protect existing old growth and to develop and maintain large, ecologically functional examples of native vegetation types via natural succession by using the passage of time to create old forest stands underrepresented in the modern landscape. Because of the many biological factors to be considered, many of which vary regionally, it is impossible to provide hard and fast rules for the creation and maintenance of DMAs. Nevertheless, we can still provide general criteria we believe are reasonable guides to design DMAs to serve as coarse filters to capture and maintain as much underrepresented landscape diversity as possible.

Obvious design considerations, other than size, include those of geographic position and shape. Although the size, shape, and location of any such new reserves are ultimately constrained by political realities, it is critical to articulate how they should be designed using biological criteria. Indeed, political realities also are a function of, among other things, well-articulated policy goals.

How Should Diversity Maintenance Areas Be Designated?

It is critical that DMAs be formally recognized in forest planning. Their boundaries should be explicitly mapped. Formal management designations should be given, equivalent to that of the Forest Service category MA-8, which is used for natural areas of national significance. Management and desired future condition should be described in detail in a manner that clearly gives biodiversity concerns priority over any forms of commodity production allowed in these areas. These two factors, the formal delineation of DMAs from other forest management zones and the express dominance of biodiversity protection within these zones, are necessary responses to the shortcomings of other models, such as New Perspectives, where biodiversity concerns are forced to compete with other multiple uses within each and every local project.

The designation of forest zones in which biodiversity concerns take precedence sidesteps specious and unwarranted conflicts over multiple use at local levels, which inflame conflict and are often resolved in favor of politically dominant user groups. Indeed, timber and game concerns already constitute de facto dominant use over large portions of eastern and western public forests. The formal designation of DMAs would change this historical pattern by providing for a greater proportion of public forest lands in which biodiversity concerns need not defer to those of timber and game interests.

The DUZ Model speaks to the needs of these groups, however, by allowing increased commodity production on other forest zones (subject, of course, to the ability of these zones to sustainably produce commodities). The establishment of DMAs, and of complementary areas where game or timber production might be intensified without loss of important elements of biodiversity at the landscape scale, can be planned for in concert. Their interdependent contributions of commodity and noncommodity benefits can only be evaluated in a sensible way at forestwide, statewide, and regional scales.

How Large Should Diversity Maintenance Areas Be?

Assuming that forest planners accept the need for broad-scale planning and will consider management alternatives that segregate lands dedicated to consumptive uses from those on which biodiversity maintenance is foremost, how large do these areas need to be and how should they be arranged in the landscape? These questions are central to conservation biology and were discussed extensively in Chapters 4–6. Biologists approaching these questions attempt to determine what minimal areas and optimal configurations might allow species sensitive to habitat fragmentation or other area, edge, or isolation effects to persist. Recall that many species are composed of multiple, short-lived subpopulations, which combined constitute a metapopulation, and that persistence of such metapopulations often depends on how large and connected they are. Many species also depend on some characteristic type and frequency of disturbance, leading to the concept of minimum critical (or dynamic) area (Chapter 4).

Patterns of species ranges and the degree of endemism also clearly differ across regions. Thus, the question of scale will need to be addressed separately and with independent expertise in each region of the country. Regions also differ widely in the degree to which representative habitats are already protected via existing parks, wilderness areas, and other biotic reserves. Gap analysis (Scott et al. 1991; Chapter 7) attempts to give direction to these planning efforts by identifying particular habitats or community types that are, as yet, still inadequately protected in·a region. In the absence of detailed information regarding smaller elements of diversity (Chapter 2), such coarse filter approaches must play a central role in protecting diversity (Noss 1987b; Hunter et al. 1988; Franklin 1993). Thus, our proposed Diversity Maintenance Areas represent a first line of defense against biodiversity losses—an attempt to efficiently protect species "whole-sale" rather than "retail."

Evidence suggests that the majority of reserved lands in eastern forests are too small to remain unaffected by known edge, area, and isolation effects (Chapters 4–6). The primary aim of the DUZ Model is to address this lack of large, reserved areas in the modern landscape. Long-term planning should include reserved areas at least an order of magnitude greater than those now seen in most eastern states in order to provide habitat blocks in which natural ecological conditions predominate.

Detractors of this proposition have countered on three main grounds. First, they were unconvinced that edge effects in particular, but also area and isolation effects, could operate at a landscape-level scale. However un-orthodox these ideas may have appeared a decade ago (Janzen 1983, 1986), they are now widely accepted by scientists (Crow et al. 1993). Second, some optimistic managers believe that they will be able to identify and mitigate pernicious ecological effects through active management not only within ex-isting reserves but throughout the landscape. We considered the logical con-sequences of this thinking in earlier chapters and find it unrealistic for both scientific and practical reasons. Finally, a few detractors correctly argue that even the largest possible reserves will be too small to protect against all known and expected edge, area, and isolation effects. We agree that what we propose may be overly conservative from a conservation perspective (cf. Noss and Cooperrider 1994) but point out that even a 10-fold increase in reserve size (to around 20 or 25% of the public forest lands in the East) would produce enormous benefits relative to biodiversity, especially if incorporated into a broad network of reserves. This proposal parallels one made for a Federal Livestock Exclosure (FLEX) system under which 20% of each parcel of fed-eral land leased to a livestock grower would be fenced and protected as a per-manently ungrazed reserve (Bock et al. 1993). Similarly, Noss (1993) pro-posed that 23.4% of the Oregon Coast Range should be designated as Class I Reserves, a classification fairly close to the DMAs proposed here.

In view of the known and predicted area, edge, and isolation effects re-viewed in earlier chapters (Table 11-1), DMAs in eastern forests should be at least 20,000 hectares in size, though preferably larger, and designed according to accepted principles of conservation biology.

Where Should Diversity Maintenance Areas Be Located?

There is no simple answer to this question, involving as it does factors such as land tenure, distribution of currently reserved lands, and the present and

Table 11-1
Minimum Size Estimates for DMAs in Northern Wisconsin

Biological criteria	Recommendations
Patch dynamics (Chapter 4)	To include a Minimum Dynamic Area within which 25% or less of the vegetation is recovering from disturbance; given periodic windthrows of 3800 hectares, a reserve of 15,000 hectares is needed.
Edge effects (Chapter 5)	An average minimum distance from edge to center of at least 8 kilometers is necessary to offset edge effects due to artificially high deer population densities in the surrounding landscape; for a circular old-growth reserve, the minimum area will be 20,000 hectares.
Area and isolation effects (Chapter 6)	Areas restored to old growth should be as large and continuous as possible.

future distribution of ecological elements such as habitats, the soils on which they grow, and the creatures and processes contained within (Hunter et al. 1988). Two rules of thumb apply, however: (1) Propose DMAs where they stand at least some chance of becoming reality and (2) try to capture as many elements of the forest landscape as will benefit from being surrounded by mature- and old-growth forest habitats.

Historically, lands with high timber value have proved least available for conservation purposes. While economic value should clearly play a role in designating which areas should and should not be devoted to commercial timber production, conservation values also need to be taken into account. Conflicts in land management are predictably most intense on economically valuable lands; yet because many elements of diversity depend on these forest habitats, reasonable and carefully designed sets of areas should be allocated to protect them.

Land tenure

With few exceptions, the creation of new, large biodiversity reserves in eastern landscapes will take place on public lands, including national, state, and county forest lands. National Forests are the most likely place to promote the DUZ Model, because of their large size, their multiple-use orientation, and the legal mandate that integrated, long-term management plans be prepared (Table 11-2). Other opportunities may arise as well. Many state forests will prepare or revise management plans during the next decade. These forests are often of significant size or are contiguous with National

Table 11-2
On Which Forest Ownerships Should DMAs Be Established?

Landowner group	Factors
National Forests	Managed for wide variety of uses Mandates to conserve diversity Large, contiguous ownership In-house biological expertise Hierarchical management
The Nature Conservancy	Mandate to conserve diversity In-house biological expertise Ownership often small and scattered
State forests	Managed for a wide variety of uses Few mandated to conserve diversity
County forests	Managed for a wide variety of uses No mandate to conserve diversity Difficult to organize and administer
Private woodland owners	Managed for a wide variety of uses No mandate to conserve diversity Ownerships small and scattered Difficult to organize and administer Private property rights predominate

Forest acreage, suggesting that they should be carefully considered when designing reserve and commodity-production networks at the landscape level.

In siting DMAs within public lands such as National Forests, managers should favor areas with few or compatible private inholdings to avoid possible conflicts in management objectives. Inholdings of private vacation cabins in a forest with long-term acquisition goals, for example, are probably compatible with DMA goals, while commercial timber inholdings may not be. Although individual National Forests have the legal requirement to protect all elements of diversity within their borders (Chapter 12), it may often be ecologically more efficient and sensible to plan DMAs in conjunction with surrounding land owners. In such cases, National Forest staff should take the lead to promote integrated planning. Such efforts are more likely to succeed if other landowners reduce local economic impacts on timber buyers or participate in biodiversity protection efforts. Many private landowners in or near National Forests are likely to embrace such management goals if doing so enhances the recreational and aesthetic values of their own lands. Easements and similar legal agreements should prove useful here.

Relationship to other reserved lands

To the degree possible, DMAs should be positioned to include forested reserve areas such as Congressionally mandated Wilderness Areas, Forest Service Research Natural Areas, state Natural Areas (as in Wisconsin), and other administratively designated reserve lands. Over time, the establishment of large blocks of mature forest will favor the development of natural disturbance regimes and other interior forest ecological dynamics within these reserved areas.

Special consideration should be given to other regional proposals for large biological reserves, particularly those carried out under the Wildlands Project rubric (Cenozoic Society 1992) or designed to allow migration in response to predicted climatic changes (Hunter et al. 1988). Unlike our proposal for DMAs, the reserves proposed by members of the Wildlands Project are directly justified, in part, by their recreational and spiritual benefits. However, both approaches apply similar scientific principles and could complement each other's biodiversity goals.

Inclusion of biological legacy at all scales

Inclusion of previously reserved areas represents only one consideration in DMA placement. Other important but less conspicuous elements of biodiversity should also be considered. Many small- and medium-scale biological elements that would benefit from inclusion within DMAs lie outside current reserves, including many small, remnant old-growth stands of eastern forests (Fig. 1-2). All available information about the distribution of such stands should be used to ensure that these stands are included, to the degree practical, within proposed DMAs. To understand the rationale for this, we briefly revisit the New Forestry Model.

One of the fundamental considerations of Jerry Franklin's New Forestry is the continuity of "biological legacy" on harvested sites (Chapter 10). In Franklin's terms, logging operations designed to maintain biological legacy do so by leaving groves of living trees and shrubs, scattered dead "snag" trees, abundant downed woody debris (rotten logs, fallen tree crowns), and relatively intact humus and soil layers. Such conditions provide some continuity of habitats for the myriad biological inhabitants of these sites. Maintaining bits and pieces of the microhabitats found in undisturbed forests helps avoid the local extinction or severe depression of populations of species dependent on these microhabitats. Regeneration after logging can then take place via the populations of these organisms already present on site. We expect the species that survive these massive but temporary site disturbances to subsequently move from their microhabitat islands into the surrounding logged habitat as ecological conditions permit. Some species will spread relatively quickly, while others will wait until the required mature or old-growth conditions are reached. It also is inevitable that some will be lost before the surrounding forest matrix returns to the old-forest conditions to which they are adapted. Yet, compared to the much smaller on-site biological legacies left by traditional logging practices, Franklin's New Forestry unquestionably favors the retention of both known and unknown components of the biota.

These provisions to maintain biological legacy were designed to retain biodiversity within timber stands. This technology may be especially important for some of the remaining tracts of western old growth where a political mandate exists for logging. We should acknowledge, however, that such mitigation efforts still represent experiments and so deserve careful subsequent monitoring to assess their success.

In forests of the eastern United States, a similar but far larger experiment has already been performed for us during the last two centuries. At a landscape scale, our biological legacy is a scattered and meager distribution of old-growth and mature forest stands (when forest habitats alone are considered). Even in Wisconsin, whose forests were devastated by commercial logging around the turn of the century, examination of timber compartment maps, aerial photographs, discussions with Agency staff, unpublished manuscripts, and our own fieldwork reveals the presence of many scattered old-growth stands (cf. Davis 1993). Like Franklin's microhabitat patches, macrohabitats of old forest scattered across the eastern United States also contain species poorly represented in the surrounding forest matrix. Although less diverse in species of some taxonomic groups than the surrounding landscape (e.g., vascular plants—Middletown and Merriam 1985), these remnant stands contain species, relationships between species, and ecological processes rare elsewhere in the region. Hence, they should be identified and included within proposed DMA boundaries whenever possible so as to maintain uncommon biological lineages and provide inocula to restore old growth.

Such stands can often be identified using existing databases, including GIS systems (Chapter 7). The pending development of some form of widespread gap analysis in eastern forests will provide even higher-quality information on the existence and distribution of old-growth and old forest stands. For DMA design, all available sources of data should be examined to determine the distribution of old forest stands for each native forest type (e.g., Leopold et al. 1988; Rusterholz 1989; Tyrrell 1992; Devall and Ramp 1992; Brown 1993; Davis 1993; Nowacki and Trianowsky 1993; Stutz 1993).

Active debate continues over technical definitions of old growth in eastern forests (Juday 1988; Barnes 1989; Hunter 1989, 1991; Martin 1992). As a rule of thumb, any forest stand in the eastern U.S. with a canopy at least a century old should be considered for inclusion in a DMA or possible reserve network on the basis of the biological legacy it may contain. Never-cut "virgin" stands are likely to retain the greatest biological legacy; but other stands, regenerated when old growth was still relatively common, may carry a substantial legacy.

In addition to stand age, the scarcity of native forest types should also be considered in designing DMAs and reserve networks. Eastern landscapes especially lack old growth of commercially valuable, long-lived, late successional species. Old growth of noncommercial species (e.g., old but tiny black spruce in bogs) or of short-lived, early successional species (e.g., 50-year-old "old-growth" aspen stands) is important to consider but typically less of a conservation priority because it is more common within the region. Caution should be practiced to ensure that such stands are not "substituted" for mature and old-growth stands of late successional species in management proposals.

So far two kinds of coarse filters have been considered here: currently existing reserved lands and old-growth or old forest stands identifiable with current technology. To position DMAs exactly, one should further consider locality data for unique elements of biodiversity (Franklin 1993). Although incomplete, current databases often include locations for rare plants and animals that would benefit by inclusion in reserved blocks of forest habitat. It's a judgement call: Although DMAs can best be used to protect habitats rather than as a means to rescue individual species, opportunities may arise in which slight alterations in DMA boundaries could offer protection to such species or sites. However, landscape-scale reserve design cannot and should not be based primarily on data nonsystematically gathered on a species-by-species basis. In practical terms, we will never have all the data we desire to make decisions about the optimum location for reserves. We can only utilize all sources of information available to us, starting with data at the largest scale and working to as small a scale as possible.

What Shape Should Diversity Maintenance Areas Have?

Many animals of northern climes are more massive than their southern counterparts, which minimizes their contact with the environment in an important way. By reducing the ratio of surface area to internal volume of their bodies, they decrease heat loss to what is seasonally a thermally hostile environment. Somewhat analogous is the issue of "body shape" for DMAs.

Considerations of edge, area, and isolation, discussed in the early chapters, all suggest DMA designs should be are as blocky or rotund as possible. The more nearly circular the DMA, the lower its edge to interior area ratio and the less likely it is to be influenced by disturbances originating from the surrounding environment.

In practical terms, this means that one should not propose relatively narrow strips of land (less than a few kilometers wide) as DMAs unless there are special reasons to expect that edge effects may be buffered. If such strips are designated as corridors between DMAs and other reserves, their area should not be counted as forest interior when assessing the biodiversity benefits for a given reserve proposal.

The benefits of corridors must be assessed on a case-by-case basis (Csuti 1991; Simberloff et al. 1992; Chapter 6). Corridors should be used to connect DMAs or other kinds of reserves when they will clearly enhance the survival of target organisms. In cases where no such organisms are identified, potential land designations might better be "spent" by increasing the size and circularity of individual DMAs.

In summary, DMA design should take into account the following factors:

1. The distribution of lands where the creation of DMAs may be feasible, especially National Forests and other large, contiguous blocks of public lands

2. The current distribution of other reserved lands, including state and federal wilderness areas, Research Natural Areas, state-owned scientific areas and Nature Conservancy holdings

3. The distribution of lands with low densities of roads, permanent openings, and private inholdings

4. Proposals for other new reserves initiated under a broader charter, such as those of the Wildlands Project and other priority habitat mapping initiatives

5. The distribution of remnant old-growth and old forest stands of all forest vegetation types, especially those once common but now poorly represented on a regional basis

6. Any and all available information on the distribution and occurrence of individual rare species and processes indigenous to forest habitats

How Should Diversity Maintenance Areas Be Managed?

We have discussed the many ways that small size and deleterious edge effects conspire to alter the ecological dynamics of nature reserves, eroding their biological legacy. If our primary goal in establishing DMAs is to promote natural ecological processes to the maximum degree possible, then it follows that we should manipulate these reserves as little as possible.

Most land managers have a strong sense of responsibility and want to take good care of forest lands in their charge. Active efforts to boost populations of one or a few valued target species often bring tangible, satisfying results and a feeling that something positive is being accomplished through direct action. There is also a certain power in realizing that by "tweaking" part of an ecosystem (e.g., removing a forest canopy, planting clover in logging roads, etc.) we can cause many attendant results. In the case of DMA management, however, we have only a rudimentary understanding of what those desirable results might be. We can only specify the desired outcome in a general way—like saying that we want a car "to run well" without knowing much about its internal workings. Since we cannot yet specify the parameters of a "healthy" and "sustainable" ecological interaction among a nitrogen-fixing *Lobaria* lichen, the *Nostoc* green alga living inside of it, the bark chemistry of the century-old yellow birch on which it is growing, the populations of phytophagous mites that eat the thallus of the lichen, and the effect of the nitrogenous compounds contributed by the lichen to the soil fungi, it is difficult to magage actively for this interaction, regardless of our intentions. We simply don't have the information necessary to take on the job in a fully responsible way.

To translate skills at managing populations of single species into management of groups of species remains a major challenge for wildlife managers. The degree to which this is a technical problem, versus a conceptual problem, is not clear. Recall that efforts used to boost populations of particular indicator species without providing for other residents of those habitats can mislead us into believing all elements of biodiversity are secure (Chapter 7). Similar problems have arisen in Wisconsin when active management efforts have been directed at maximizing populations of single, large mammal species, including wolves and deer, sometimes to the detriment of other ecological values (Chapter 13).

The management goal of the DMAs, in contrast, is not to maximize populations of wolves, hemlocks, microscopic collembolan insects, or of any individual species. It is to create areas containing the dynamic ecological processes to which mature and old-growth forest species are adapted, and within which these species can interact and continue to evolve without severe regimes of artificial selection. The goal is wildness, rather than an exact, static recreation of presettlement conditions (Noss 1991a). Even if these conditions were known with some accuracy, it is unrealistic to assume that they could be recreated and efficiently maintained.

If the goal for DMAs is to reestablish natural ecological conditions and disturbance regimes, it would be most cost-effective and least presumptuous to make our restoration efforts passive and patient. Remnant old-growth and old forest stands included within DMA boundaries should be left unmanipulated, as they are needed to act as inocula for restoration. Biologist Daniel Janzen has long talked about remnant stands of dry lowland neotropical forest as "biotic debris, the restorationist's biotic tools" (Janzen 1988). What we propose here is analogous.

In Wisconsin, at least, these remnant old-growth stands are usually surrounded by younger woodlands. It is likely that the gradual aging of these woodlands adjacent to remnant old-growth stands will, over the course of decades and centuries, provide new habitats into which many species in the old-growth stands will gradually disperse and colonize. Unfortunately, we know rather little about how and how quickly these organisms colonize new habitats. While we fully support research to answer these questions, it would be inappropriate to presume that we now have enough information to automatically prescribe artificial introductions. Until research results justify more active restoration projects, it is most prudent to restrict ourselves to ameliorating obvious edge effects and eliminating exotics within areas designated as DMAs (Temple 1990).

Those who are anxious to begin active restoration should direct their efforts into careful experiments in forest zones outside of the DMAs. The results of these experiments, if carefully monitored and compared with unmanipulated control areas in the DMAs, can yield information on the efficacy of active restoration efforts (Chapter 7). In Wisconsin and Michigan, a study is now underway to compare the structure and composition of periodically harvested, uneven-aged stands with unmanipulated old-growth stands on sites with similar soil and landform characteristics (C. Lorimer, pers. comm.). Through such studies, we can begin to discover to what degree managing for some aspects of old growth promotes other old-growth attributes according to a broad spectrum of resident organisms and ecological dynamics. We will have more satisfying answers sometime in the next century.

There is no simple response to this or any way to be a perfectly biocentric manager, and tough decisions are required to counteract these internal and external threats to diversity (Diamond 1992; Alverson and Waller 1993). In general, however, we recommend an interim hands-off management strategy in DMAs and further research into old-growth components and processes.

Compatible Multiple Uses within the Diversity Maintenance Areas

Just as it is critical to spell out as precisely as possible the desired future condition of DMAs, it is important to specify those types of activities thought to be compatible or incompatible with the dominant-use objectives of the DMAs. Any and all uses compatible with the protection and enhancement of biodiversity should be allowed in these zones. Activities for which there are reasonable doubts, or factual evidence against compatibility with DMA objectives, should be excluded until their compatibility can be established. This is quite different from the current situation, at least on National Forest lands, where the onus is on those concerned with the native biota to prove in each and every local project decision the negative consequences that forest manipulation will produce in the future.

Below, we review what we consider compatible and incompatible forest uses within DMAs designed for Wisconsin's two National Forests. While regions differ, these same categories of uses will be of concern in most other eastern forests.

Harvest of forest products within the DMAs

Commercial timber production in the form of clearcuts, strip-cuts, seedtree cuts, and shelterwood cuts creates patches of young, "disturbed" forest habitat. In a general way, these disturbances appear akin to natural disturbances such as windthrows. They differ in several respects, however, such as by their much greater spatial and temporal frequency of occurrence (Chapter 4). Our aim, in proposing DMAs, is to provide large areas with the natural disturbance regimes resulting from windthrow and fire. For example, fires would be allowed unless they posed a threat to adjacent areas, much like existing Wilderness Area fire policy.

In contrast, commercial logging and artificially created wildlife openings should be excluded from DMAs because they artificially increase the amount of young habitat and the negative edge effects associated with that type of habitat. On the other hand, very small-scale extraction of wood and other forest products should be considered in DMAs as long as there is evidence that they do not compromise long-term goals. In Wisconsin, we proposed that noncommercial extraction of firewood for local residents be continued permanently on a permit basis. Likewise, the harvest of balsam fir boughs (which can be sold to local industries that manufacture wreaths) should be continued because the overall impact on these trees, which are common, is minor. Other activities, such as berry picking and harvest of other common wild plants and fungi, should be considered favorably. Western forests are now assessing the economic and ecological significance of such nontraditional harvests, and similar efforts will likely follow in the East.

Perhaps more controversial than the small-scale extraction of plant products from these areas is hunting. In the Wisconsin DMAs, we argued that hunting should not only be permitted but should actually be increased for deer. Correspondingly, we asked for a cessation of game stocking and other active game management within the DMAs, as these are largely edge-loving species already overrepresented at the landscape level. This overrepresentation has resulted from habitat manipulation and from the reduction of natural predators, wolves and cougar in particular (though black bears continue to be an important predator on young deer). Thus, within the DMAs, it is beneficial to address this superabundance of herbivores and other edge-loving game through increased hunting (Garrott et al. 1993). Such hunting would have to ensure that nontarget mammals such as wolves and moose were not shot in the process. Over decades, as the average age of the vegetation increases, less favorable habitat will be available for edge-loving game species, decreasing the quantity of hunting opportunities although not necessarily the quality. Efforts to increase game abundance would be redirected to other forest zones.

Recreational use and motorized vehicles within DMAs

Moderate levels of nonextractive recreation compatible with the management goals of DMAs should be encouraged, including hiking, camping, skiing, mountain bicycling, and nature study. Large-scale expansion of any of these activities, or expanding motorized or extractive forms of recreation within the DMAs, should be delayed until their possibly undesirable ecological effects can be assessed.

In contrast to Congressionally mandated Wilderness Areas, and the mega-reserves promoted by the Wildlands Project (Cenozoic Society 1992), recreational and spiritual experiences are not primary justifications for DMAs. Although these benefits could likely occur within DMAs, their production would not be a primary management goal or a dominant use. Unlike these other kinds of reserves, we sought to include or exclude motorized vehicles solely on the basis of their impact on biodiversity.

Our Wisconsin proposals called for maintaining existing main roads within DMAs but asked that the smaller (Class D and logging) roads be gradually abandoned. The objective was to reduce the amount of edge habitat and forage associated with these roads and to reduce overall road density to below one mile of road for each square mile of forest. The latter value is the road density below which areas can function well as wolf habitat, since people are more likely to hit or shoot wolves at higher road densities (Thiel 1985; Mech et al. 1988).

Snowmobiles are common in our region, and we agreed that existing trails and access be maintained. Although one can imagine conflicts arising from harassment of wildlife during winter months, we have not yet seen evidence that this occurs in any way contrary to the objectives of the DMAs. Such use should be carefully monitored to verify that responsible snowmobile use is compatible with wintering populations of large mammals such as wolves, moose, bear, and cougar.

Noss's criteria for Class I biodiversity reserves in the Oregon Coast Range differ slightly from the management criteria proposed here for DMAs in that he specifically excludes any harvesting of natural products, motorized vehicles, and mountain bikes (Noss 1993). His list of management guidelines for Class I, and for the less restrictive Class II and Multiple-Use Buffer Zones, should also be considered by conservationists evaluating compatible and incompatible uses of particular reserves in eastern forests.

Managing Non-DMA Forest Zones for Other Dominant Uses

A fundamental tenet of the DUZ Model is that dominant uses other than biodiversity management be addressed within other zones of multiple-use lands. For example, if one assumes that the ASQ (maximum allowable sale quantity), or the actual timber targets that are set as a result of budgeting, for any given National Forest will not be reduced in the next few decades, then one must compensate for timber sales displaced by areas to be designated as DMAs.

We do not concede the need or wisdom of current ASQ or timber target levels. For example, we envision that forestry will go through the change witnessed in the last decade in the electric power industry: the "old days" in which growth of base load was the mark of good management are being replaced with state regulatory orders that companies invest in demand side management projects and research. There are policy and legislative economic incentives that hamper such a change in forest management (Chapter 12), and these should be addressed as an integral part of reforming forest management.

We also reject the simplistic notion that there is an inevitable choice between sacrificing U.S. biodiversity and that of other wood-producing nations (Bowyer 1992). There is a raft of inquiries into recycling, reuse, and other sources of demand side reduction that should be explored before deciding that biodiversity must go.

But if, despite all these best efforts (which have not yet been made in any serious way by our society), it remains necessary to intensify harvesting in some areas in order to preserve DMAs, what would be the result? If roughly 20% of a forested area was newly devoted to DMAs, then roughly 80% of the area previously available for timber production would have to produce 100% of the timber, an increase in silvicultural intensity of about 25% on those acres. Actual values would depend on what percentage of the overall timber base (lands specifically devoted to producing timber) is included within DMAs. Here we assume that, on average, DMA and non-DMA lands would contain similar densities of timber.

Such a proposal raises many questions for which we lack solid answers. Foremost is the question of whether these DMA zones can support sustainable forestry at 125% of the current levels. Certain lines of evidence suggest that they can, though further evaluation is needed. The Nicolet National Forest Plan stated that the level of timber harvest could range to more than three times the current harvest level (of 74 million board feet in 1984), to about 222 million board feet per year. Harvests proposed for the first decade (1986–1995) ranged from 72 to 97 million board feet per year, the higher volume representing the Agency's preferred alternative, which was 131% of the volume harvested in 1984 and putatively only about 60% of what the forest was capable of sustainably producing. Harvests proposed for the fifth decade ranged from 120 to 186% of the 1984 volume, the higher figure representing the Agency's preferred alternative (USDA Forest Service 1986b). European forestry provides comparable outputs on a sustainable basis but does so by more intensive forestry efforts than are common in this country.

A somewhat less analogous but instructive example is that of New Zealand, where the equivalent of our Forest Service was cleft into separate agencies in the mid-1980s (Kirkland 1988). One branch manages production forest lands, largely plantations of nonnative Monterey pines, which produce 95% of the commercial timber yield on about 15% of the overall forest acreage, enough to make New Zealand a timber-exporting country. Natural forests are used only incidentally for timber production, producing less than 5% of the timber on 85% of the forest acreage, allowing them to be far less affected than ours by silvicultural activities.

We believe that we can do a better job, both ecologically and economically, by specializing management within large zones of our eastern forests. A 20,000-hectare DMA plus 80,000 hectares of forest land with slightly intensified silviculture would be superior, from the standpoint of biological diversity, to 100,000 hectares of land on which silviculture is practiced and biodiversity is provided for primarily via mitigation. The latter scenario might produce larger populations of creatures adapted to intermediate disturbance regimes, but it is much less likely to protect sensitive elements of biodiversity that now suffer and are poorly represented at a landscape level. The aim should be to adequately provide for truly unique elements of diversity and ecological dynamics somewhere, not just to do a marginally better job everywhere.

The DUZ Model was proposed out of concern for components of biodiversity unique at the regional level. Aside from these biological considerations, there may be economically compelling reasons as well for adopting the model, perhaps at a scale above that of individual National Forests. Serious economic analyses of the DUZ Model are only now underway (L. Travis, pers. comm.). Many questions remain unanswered: Is more intensive forestry in some zones often more cost efficient at a forest-level scale? What will be the effect on overall hunter visitation if some areas have fewer (but some larger) deer while the deer populations increase in other forest zones? How will recreational use of public forest lands change if large, old-growth stands are allowed to develop?

Recent changes in national politics suggest that timber harvests will decline on many National Forests. How this will affect the relative production of timber on other public and private lands remains to be seen. Ironically, such changes may create *de facto* zoning at a large scale by transferring regional timber production back to county, state, corporate, and private lands. If Agency timber harvests decline, the DUZ Model will become easier to institute on National Forest lands.

Summary

The Dominant Use Zoning Model assumes that the most conservative course of action, from a biodiversity standpoint, is to carefully designate and passively create large, wild forests where human influence is kept to a minimum. These large areas are not proposed as substitutes for the current system of small and medium-sized biological reserves, but as a necessary addition. Nor do we suggest that remnant habitat islands of old growth outside of the DMAs should be cut. The management of these DMAs would not be heavily dependent on tenuous annual funding for monitoring and mitigation efforts as are models that purport to protect biodiversity in areas simultaneously subject to logging. Monitoring would be useful, however, to detect the effects of deleterious edge effects and to compare long-term successional changes in the newly designated DMAs with other lands subject to direct manipulation. Indeed, such information would allow us to directly compare DMAs to experimental New Perspectives and active restoration projects elsewhere in the forest landscape.

We encourage independent cost-benefit analyses of the effects of the DUZ Model in terms of timber, game, recreation, and nontraditional commodity production. It is our belief that this model can produce superior results, or at least outputs equaling those at present, in most or all of these categories, but serious analyses are desperately needed. Only with this information can we accurately evaluate this model within a broader socioeconomic context.

PART IV

Toward a New Diversity Policy (and Twenty-First Century Old Growth)

IN THE PRECEDING CHAPTERS, we have described the biological costs of forest fragmentation and proposed as a response the minimization of human-caused disturbances in large blocks of naturally maturing forest. This proposal would bring portions of our forests full circle from the old-growth ecosystems still largely dominant at the time of European settlement. Forest policy in this country began with Gifford Pinchot's call to halt outright destruction, moved to the artificial boosting of populations of desired tree and game species through dramatic manipulation of the landscape, tried to stem continuing species loss through emergency single-species rescue operations, and now must progress to a conscious choice of wild forest conditions in biologically significant locations and quantities.

Each of the developmental stages in American forest policy remains highly relevant to the current practice of forest management. In parts of western and southcentral United States, elementary conservation is still not fully accepted. Many areas remain subject to outright destruction through extensive clearcutting causing damage Pinchot would have deplored: virtually irreversible habitat loss, soil erosion, and stream siltation. Basic conservation is practiced in most of our public lands, but management typically takes the form of intensive manipulation of tree and game populations selected for increase. This was the dominant paradigm in the Land and Resource Management Plans adopted by the Forest Service in the mid- to late 1980s under the National Forest Management Act (NFMA). The Endangered Species Act (ESA) remains the most powerful weapon against species loss, but it has focused on saving individual species near extinction and has in some cases even been detrimental to ecosystem protection by manipulating environments for the benefit of single species in ignorance of the structure, function, and innumerable components of the ecosystem needed to support the full web of life.

While conservation (as simple restraint), intensive management, and localized mitigative habitat engineering will remain integral parts of forest management in the foreseeable future, they are plainly inadequate to the task of protecting biodiversity. The list of threatened and endangered species continues to grow, and, with few exceptions, the amount of serviceable habitat continues to shrink. It is now time to recognize and act upon the idea that the indispensable core of future efforts to protect the native biological diversity of our forests must be a return to natural patterns and forms of landscape disturbance, eventually leading to old-growth forest conditions in substantial portions of the landscape. Compromises from this goal must be clearly stated as such and recognized as something less than ecosystem protection. In some areas, our society no longer may be able to afford such compromises.

Our task in the remaining chapters is to translate this conclusion into a new, affirmative diversity policy for our National Forests. While Chapter 8 reviewed the evolution of forest management from the standpoint of the effects those policies wrought upon the forests themselves, Chapter 12 examines four issues in the history of forest management for the role they might play as building blocks, or stumbling blocks, in the task of constructing a new diversity policy. We will also review prospects for diversity protection arising from the three major statutes adopted in the last 20 years governing the obligations of the Forest Service to protect the environment.

Chapter 13 examines the specific cases of the dispute over biodiversity in the two Wisconsin National Forests. Although not intervening quite as directly in the violence to our public lands as some activists recommend (Foreman 1991a), we have entered into the policy arena, filing administrative appeals and litigation to achieve better protection of diversity. These cases provide valuable insights aiding our synthesis of science, management models, public policy, and law into recommendations for a new diversity policy (Chapter 14).

Our decision to devote this concluding section to the practical problems of implementation through policy and law should not diminish the ethical implications of our conclusions about the direction of a new forest policy. On the contrary, the lessons of conservation biology form a singularly compelling refutation of resourcism as exemplified by Pinchot, "wise use," and the post-War Forest Service. Leopold fully made this case more than four decades ago, and yet even later in life remained gloomy about the prospects for a new ethical system in favor of the land's own biological self-determination (Meine 1988; Oelschlaeger 1991).

We view our suggestions as the implementation of an ethical choice, not in lieu of one—and not only for the proper care of the land but also for the necessary role of science in these latter days (i.e., 43 years after *A Sand County Almanac*). In describing Leopold, Oelschlaeger tells us

> The land ethic is clearly a revolutionary departure from the stance of modern science, which permits the scientist to describe but never prescribe. Yet Leopold was convinced that ecological knowledge must be brought to bear upon prescriptive questions concerning the interrelations between *Homo sapiens* and the larger biotic community. (Oelschlaeger 1991, p. 206)

It is in this spirit that we will venture a few prescriptions of our own.

12

Sources of a New Diversity Policy for the National Forests

The need to establish and implement an appropriate policy for the protection of biodiversity is of immediate importance for the National Forests. For several reasons, these lands are at the forefront in the development of our domestic policy for biodiversity. Most visibly to the public, the National Forests of the Pacific Northwest contain the last remaining old-growth forested ecosystems in the contiguous states. Less apparent to the public is the fact that the Forest Service has been asked by Congress to articulate the first national policies and management strategies for diversity. The National Forests are the only public lands subject to an express Congressional mandate to "provide for diversity of plant and animal communities" [16 United States Code (U.S.C.) § 1604(g)(3)(B)]. And beyond the Pacific Northwest, the National Forests are large enough and biologically rich enough to provide real protections against further losses in diversity.

The tension between protection of the land and timber outputs (Chapter 8) has been a central National Forest policy issue at least since the Second World War. However, it is only in the last few years that stemming widespread losses in biodiversity has been acknowledged as a major concern for forest management. This more recent focus is due in part to the fact that a great deal of ecological understanding progressed at an accelerating rate in the scientific community largely independent from the process of internal Forest Service policy development. When the National Forest Management Act (NFMA) was enacted in October 1976, it allowed existing timber management plans to remain in effect and called for the development of new forest-wide plans, which the Forest Service was to "attempt to complete by no later than September 30, 1985" [16 U.S.C. § 1604(c)]. Thus, during the 1970s and 1980s—a time of significant progress in ecological understanding—the Forest Service's management personnel worked internally to develop Forest Plans with little participation from the Research Branch of the Agency and virtually no peer review from the scientific community concerned with non-commodity conservation.

After a decade-long gestation period, the first round of Forest Plans under NFMA was issued in draft and then final form in the mid-1980s. Environmental activists were inundated in a sea of documentation and gave priority to concerns about individual endangered species, clearcutting, below-cost sales, stream siltation, road networks, pristine recreational areas, and other traditional concerns of environmental groups. These issues were more easily related to earlier environmental agendas and were thought to be more effective in fundraising and membership drives than the more abstract concept of biodiversity. Also, few activists were trained in evolutionary or conservation biology or were aware of the cumulative impacts of forest fragmentation. As the issue of biodiversity became more prominent in the late 1980s as a worldwide concern, activists gradually recast many of their complaints about National Forest management as threats to biodiversity. A new journal, *Wild Earth,* even proclaimed the birth of a "New Conservation Movement" grounded in the tenets of conservation biology as the basis for future public land planning (Foreman 1991b).

The problem of integrating meaningful protection for biological diversity into the overall management scheme of the National Forests is now squarely before the Agency and the public. If we wish to influence forest policy with the principles of the preceding chapters, where best to begin? It can be argued persuasively that the most effective strategy would be to set the scientific arguments aside while focusing one's efforts on the political and economic factors driving the principal cause of landscape disturbance—timber harvesting. While it is undeniable that the levels of timber production adopted in the Forest Plans of the 1980s often preclude substantial protection for biodiversity, we will not focus on a strategy for reducing timber volumes *per se* for several reasons.

The subject of timber dominance in National Forest policy has been capably addressed elsewhere, particularly in the periodical literature of the last few years (e.g., Skow 1988; Knize 1991). The causes of timber dominance, both historically and in current policy making, have also been ably addressed, particularly the role of the budgetary process in promoting high levels of harvest (O'Toole 1988). One must recognize the historical and current dominance of timber to rationally propose any new policies for the Forest Service, yet there are dangers in approaching every issue, whether recreation, watersheds, or diversity, as primarily a matter of modifying the Agency's timber policies.

A distinct diversity policy for the National Forests must be separately stated, aside from desired timber harvest levels. Proponents of diversity protection must be able to articulate in specific terms both the magnitude and the nature of management initiatives needed to provide for diversity.

There must be a specific diversity policy ready when vague notions of the need for ecosystem protection begin to loosen timber's control of the Agency.

Second, a substantial and vigorously applied diversity component of forest policy could hasten reductions in the ubiquitous logging throughout non-wilderness lands (and threats to wilderness from the Wise Use movement). The increased visibility of biodiversity in the northern spotted owl controversy and the 1992 Earth Summit may make diversity protection a singularly effective force in resetting the balance of forest management.

Finally, the Agency has been busily crafting its own notions of appropriate policy for diversity. The Forest Service developed methods to define, evaluate, and provide for diversity for use in the first round of management plans in the mid-1980s (Chapter 13). The Agency's policy has been restated and refined in the Chief's decisions on the appeals of the two Wisconsin National Forest Plans in early 1990, in the ensuing Wisconsin litigation (Chapter 13), and in drafts of revised planning rules proposed by the Agency in early 1991. Thus, the *de facto* lead agency for developing our domestic biodiversity policy has been busily defining its policy toward biodiversity protection while many Agency watchers focused on other issues. Anyone concerned with influencing the development of diversity policy for the National Forests (and through this Agency's leadership, national policy in general) must address the issue before the Agency's own views become hardened, both institutionally and legally, in administrative interpretations, formal planning documents, appeal decisions, and perhaps amendments to substantive planning regulations.

Too often, excellent scientific analyses of threats to diversity demur on the policy changes necessary to stem those threats or naively assume that the flexibility of multiple-use principles guiding the Forest Service will readily accommodate the need for greater diversity protection (e.g., Terborgh 1989). Our experience in trying to incorporate diversity protections in the Forest Plans for the two Wisconsin National Forests made clear that there is a substantial gulf between describing ecological problems based on sound scientific information and securing changes in management to respond to those environmental impacts. To successfully bridge that gulf, we should begin by examining several major historical and legal influences in the Agency's past which are central to the Agency's current and future treatment of diversity.

Are There Roots for a Mandate to Protect Biodiversity in the Historical Development of Federal Forest Policy?

Forest policy has evolved almost entirely in the absence of concern for composition, structure, and function at the ecosystem level. When faced with the statutory mandate of NFMA to protect diversity, the Agency responded with methods and policies firmly grounded in its history and institutional culture.

Pinchot's outlook remains central to the Agency's treatment of the modern problem of biodiversity protection

The roots of Forest Service thinking are properly traced to the utilitarian philosophy of its founder, Gifford Pinchot. The basic tenets of Pinchot's outlook have been addressed in many other writings (Pinchot 1947; Udall 1963; Nash 1967; Pinkett 1970; Wilkinson and Anderson 1985). Our present task is to see how his philosophy influences the modern treatment of diversity.

Pinchot's forest was made up of a mix of tree species rather than an entire web of interdependent plants and animals. Pinchot's concern for sustainability was limited to the economic parameters of productivity and profitability in the harvest of timber. Even this narrow view of sustainability has not been fully embraced by the modern Forest Service, as evident in its unwillingness to acknowledge the unsustainable levels of logging in the old growth of the Pacific Northwest and its delay in confronting the issue of long-term soil productivity even for silvicultural purposes (Chapter 8). But these issues are within the limited bounds of Pinchot's conception of forests as trees, a view which is not concerned with the much more complex relationships and interdependencies perceived by Muir and Leopold. In Pinchot's tradition, the Agency relied heavily on data regarding the mix of tree species and age classes in its efforts to address its statutory duty to protect diversity in the first round of NFMA planning in the 1980s (Chapter 13).

Can we modernize Pinchot's notions of sustainability and forest composition by broadening them to address biodiversity? Even if adapted to a modern setting to include species, such as the red-backed vole, that affect growth of certain tree species (Maser 1989), or an expansion beyond trees to other "products" of the forest such as the cancer drug Taxol (Nicholson 1992), the Pinchot approach would still be a limited source of inspiration for a new diversity policy. Even if one were to extend Pinchot's early notion of sustainability beyond trees, his approach would favor active over passive measures (Chapters 3 and 8) and would focus on the regeneration of preferred species rather than acceptance of a species mix dictated largely by the ecological forces at play in the natural community of plants and animals. Without suggesting that diversity protection must be conducted entirely through passive restoration methods, it is still useful to recognize that Pinchot was committed in the opposite direction. His belief in active management methods, when coupled with his focus on preferred individual species, would, if applied to the sustainability of species other than trees, lead one in the direction of an aggregation of mitigation plans for species that can already be identified as rare or otherwise selected for value, rather than an integrated ecological approach.

This strong affinity for active management methods and the belief that human-devised plans for altering populations of particular species of concern can fully address the diversity problem have been the cornerstones of the Agency's approach to diversity, both in the early 1980s, when developing the first round of Forest Plans under NFMA, and in the "second-generation" diversity policy that has been developed by the Forest Service in part in response to the Wisconsin cases. In the Forest Plans for the Wisconsin National Forests, sensitive species habitat needs were addressed through standards and guidelines consisting of local mitigation plans. As an example of the Agency's response to the appeals of those plans, the Agency proposed to protect Canada yew from herbivory through fencing particular plots (Chequamegon Forest Plan Amendment #3, 5/15/91, subsequently withdrawn after appeal). The view that an aggregation of mitigation plans directed at particular species can anticipate and adequately repair all the plant and animal interconnections that constitute a habitat is a direct descendant of Pinchot's approach to forest management.

Lastly, Pinchot's aversion to the wild state of the forest—his desire to convert it to managed second growth—has also manifested itself in the Agency's treatment of the diversity problem. As Wilkinson and Anderson (1985 p. 134) explained:

> Another important element of Pinchot's policy was to convert the wild, old-growth stands to scientifically managed second-growth. Pinchot reported in 1908 that "[f]ull utilization of the productive power of the Forests . . . does not take place until after the land has been cut over in accordance with the rules of scientific forestry. The transformation from a wild to a cultivated forest must be brought about by the ax. Hence the importance of substituting, as fast as practicable, actual use for the mere hoarding of timber." (Footnote omitted)

In this language, we see that, unlike Muir, Pinchot did not acknowledge any value in wild forest conditions. Similarly, the modern Forest Service did not value wild forest conditions as a way of responding to the Agency's statutory diversity mandate. When the authors argued that regulatory language which called for diversity "at least as great as that which would be expected in a natural forest" meant restoration of natural ecological conditions, the Agency told us such natural processes were an inappropriate goal in secondary growth, managed forests (Chapter 13).

Thus three key elements of Pinchot's thinking—a view of forests as simply trees, hubris that humans could actively manipulate the forest to any desired condition, and a belief that wild (natural) conditions were antithetical to wise management—remain an often unstated but nonetheless critical foundation for the Agency's modern approach to protecting biodiversity, not only in

Forest Plans of the 1980s, but in its responses to the challenges to those plans in the 1990s.

The perennial clearcutting controversy may distract us from broader biological processes and losses

The public's adverse reaction to the practice of clearcutting and other forms of even-aged management has had a prominent place in public debate over Forest Service management ever since the Second World War. Selective cutting had dominated the management of many National Forests at their inception. But after the war, the greater efficiency of clearcuts for extracting timber and the amenability of cleared areas to more intensive management caused the Agency increasingly to rely on this method of logging and to defend it on "scientific" grounds (see, e.g., Chief Cliff's book *Timber Management*—Shepherd 1975).

The heavy use of clearcuts, even in steep western terrain, drew increasing criticism from a growing environmental community. Congress first addressed the practice when Senator Lee Metcalf of Montana commissioned Dean Arnold Bolle to assemble a group of independent forestry experts to evaluate cutting practices on the Bitterroot National Forest. The subsequent Bolle Report (1970) condemned clearcuts as timber mining and concluded that "Multiple use management, in fact, does not exist as the governing principle." Chief Cliff responded to the ensuing public outrage with promises to control harvesting practices more carefully and an appeal to Congress to appropriate more funds to other uses, such as recreation (Clary 1986). Viewing public objections in part as a public relations failure, the Forest Service also began to hire landscape architects to hide the scars of clearcutting behind buffers (known as "beauty" or "fool 'em strips") adjacent to public roads and watercourses (Lansky 1992).

The clearcutting issue created a landmark in federal forest policy with the decision in the *Monongahela* case [*West Virginia Div. of Izaak Walton League of America, Inc. v. Butz*, 522 F.2d 945 (4th Cir. 1975)]. A group of West Virginia citizens obtained an injunction barring clearcutting on the basis of language in the 1897 Organic Act, which allowed the harvest of only "dead, matured, or large growth of trees" that had been "marked and designated" before sale. The Fourth Circuit Court of Appeals upheld the trial court's ruling, finding that "matured" meant physiological maturity rather than economic or management maturity, as the Forest Service had urged. The appellate court reviewed the history of the Agency, finding that it had "changed from a custodian to a production agency," and declined to rewrite the Agency's enabling legislation to fit this "production agency's" view of appropriate management. Heeding the court's direction to seek relief in Congress, the industry and the

Forest Service successfully overturned the *Monongahela* decision's outright ban on clearcutting through provisions in the National Forest Management Act of 1976 (NFMA).

The high visibility of clearcuts has contributed to the prominence of this issue. The practice is so apparently destructive, and its environmental impacts so obvious, that it has earned legitimate and substantial criticism from a broad range of concerned citizens. The patchwork of clearcuts and similar even-aged harvesting methods (seed tree and shelterwood) removes virtually all biomass, cover and nesting places from an area, contributes to fragmentation, and encourages erosion, reducing or eliminating vegetative regeneration while fouling aquatic habitat. Recent evidence suggests that in some forest types, understory herbaceous plants may not recover from clearcutting until after a lengthy cycle of succession (Duffy and Meier 1992).

However, the corollaries often found in the literature that selective logging practices are the answer to sound forest management (G. Robinson 1988), or that a reduction in even-aged management will, in and of itself, substantially alleviate losses in diversity in the temperate forests, are misplaced. The smaller clearcut size of 40 acres mandated as a general rule by NFMA regulations for areas outside the Pacific Northwest and southeastern U.S. [36 C.F.R. § 219.27(d)(2)] results in a greater dispersion of cuts through a wider area of the forest, generally demands a more intensive road network, and maximizes the creation of additional edge habitat often detrimental to edge-sensitive elements of diversity (Chapter 5). Thus, hiding and dispersing clearcuts does not necessarily reduce their ultimate biological impact, but may instead be doing just the opposite.

Both the Agency and some environmental activists have treated the amount of this one harvesting practice as defining the character of the forest and its potential to sustain biodiversity, rather than seeing it as only the most glaring example among a set of landscape disturbances whose impacts should be measured not only in terms of local intensity but also in terms of forestwide extensivity. The Forest Service expressly sought to use the amount of clearcutting as the barometer of biodiversity protection during the 1992 Earth Summit in Rio de Janeiro. When President Bush was embroiled in the controversy over his refusal to sign the Biodiversity Convention resulting from the summit, the Chief of the Forest Service announced on June 4, 1992 a new policy suggesting reduced clearcutting (Robertson 1992). Aside from the substantial exceptions which were provided to allow continued use of the practice, the important message from that announcement was that the American public was led to believe that the way to improve "protection of plants and animals" (as the news media described biodiversity at the time) was to reduce clearcutting.

Many environmental activists foster this view by focusing lobbying efforts on legislation which directly links reductions in clearcutting and substitution of selective logging, on a stand-by-stand basis, with local improvements in biodiversity (e.g., Forest Biodiversity and Clearcutting Prohibition Act of 1993, H.R. 1164, 103d Congress). Of course, clearcutting foes do not assume fixed timber production levels and hope that greatly diminished clearcutting will lead to substantially reduced total levels of timber harvesting and a lower level of overall landscape disturbance. However, one cannot assume that reductions in clearcutting will lead to the creation of unfragmented reserved areas. On the contrary, the Agency's Integrated Resource Management, New Perspectives, and Ecosystem Management programs jealously guard against reductions in acreage deemed suitable for timber harvesting just as vigorously as lowered sale quantities, if not more so.

Thus, without assured reductions in timber targets and acreage deemed suitable for harvest, a continued emphasis on stand-level impacts and silvicultural cures to the threats to biodiversity reinforces the view that if we attend to species on a stand-by-stand basis, and do a better job in our "engineering" techniques, we can effectively protect biodiversity. This viewpoint hinders progress toward an appreciation of the cumulative impacts of other forms of logging and associated fragmentation, impeding development of a progressive diversity policy that encompasses not only local intensity of disturbance but also the overall extent and pattern of disturbance.

Wilderness: Can the idea of wilderness as, in part, a last biological refuge contribute to the pursuit of wild conditions as a key component in a new diversity policy?

Wilderness has played an important historical and philosophical role in the development of American thought about the natural world (Nash 1967). While the rationales for wilderness are many—recreational, aesthetic, spiritual, ecological, and biocentric—our present inquiry will focus on whether the idea of wilderness might aid in the development of a new diversity policy. Since "wild forests" are at the heart of our approach to protecting biodiversity, we must explore the idea and practice of wilderness protection in some depth.

Pinchot opposed wilderness in concept and practice. In his day, "[w]ild nature was conceived as little more than a stockpile of raw materials of no intrinsic value; only through the productive enterprise—the humanizing of the wilderness—did nature gain value" (Oelschlaeger 1991). Pinchot strongly opposed retaining old-growth stands and exempting forest lands from his principles of "scientific forestry." His opposition to the views of John Muir and the proposals for preservation of Hetch Hetchy and the Adirondacks are specific examples of this view (Nash 1967; Wilkinson and Anderson 1985).

Notwithstanding Pinchot's outlook, the Forest Service, through the efforts of Aldo Leopold and Arthur Carhart, was the first federal agency to protect land as wilderness. Leopold's proposed Gila Primitive Area was established in 1924. Although the roots of the wilderness idea trace back much farther, at least to Thoreau and Muir in this country, Leopold's initiative was the first explicit public policy choice by an individual charged with land management responsibility to leave the land as he found it (Leopold 1921; Frome 1962; Nash 1967).

Other Wilderness Areas were subsequently designated in the National Forests, and indeed the great majority are currently within the National Forests. Through the influence of Bob Marshall and a series of regulations adopted by the Agency, the notion of wilderness became a legitimate component of forest planning. Marshall saw wilderness primarily as a place of mystery and psychological renewal for humans rather than for its own biotic importance, yet his influence gained acceptance of the idea that barring active management was an appropriate way to care for the land.

Although Marshall and others carried forward the wilderness idea in practice, Leopold appreciated and articulated the biological value of wild lands decades ahead of the scientific evidence which now substantiates his views. Most commentators attribute Leopold's early thinking about wilderness to its recreational value (and, to a degree, its spiritual value) (Meine 1988; Wilkinson and Anderson 1985), although some recognize a concern for wild plants and animals (Frome 1962). At least in his early conception of wilderness, Leopold was willing to justify wilderness within the Pinchot framework, arguing that an area "premised wholly on the highest recreational use" was in accord with Pinchot's notion of the "highest use" of forestland (Leopold 1921).

But in later writing, Leopold identified the importance of wilderness as a "land laboratory" providing a "base datum of normality," with a "large value to land-science," and that in comparison to this role of wilderness areas, "recreation is not their only or even their principal utility" (Leopold 1941). He eventually progressed to a recognition of the intrinsic value of nature, moving from a viewpoint based in the manipulative, "imperial" ecology (an ecology founded on Pinchot's management mentality) toward a "radical, Muir-like preservationist philosophy" (Oelschlaeger 1991). As if needing no explanatory connection, Oelschlaeger treats Leopold's work in ecology as a central pillar of the philosophy of wilderness.

On the more practical level of a natural historian, Leopold saw the effects that wilderness, or lack of it, would have on the biota. In typically prescient fashion, he noted a "growing realization that only wolves and lions can insure the forest against destruction by deer" (Leopold 1939), recognized the importance of "the underground community of bacteria, molds, fungi, insects, and burrowing mammals which constitute half the environment of a tree" (Leopold 1942), and saw that "[m]any animal species, for reasons unknown, do not seem to thrive as detached islands of population" (Leopold 1953). These were harbingers of modern conservation biology: the problems of deer herbivory as a prevalent edge effect (Alverson et al. 1988), the operation of mycorrhizal fungi and small mammals in tree growth (cf. Maser 1989), and the principles of island biogeography (MacArthur and Wilson 1967). Nash tells us that more contemporary figures, "following in the intellectual footsteps of Aldo Leopold," recognized the importance of wilderness "as a reservoir of normal ecological processes and of a diversity of genetic raw material" (Nash 1967). Nash identifies former Sierra Club leader David Brower and renowned sociobiologist Edward O. Wilson as among those recognizing wilderness as "an environment where the full complement of life forms remained intact" and as "a region of biological complexity" (Nash 1967).

More recently, the connection between wild areas and diversity has been bluntly stated by Noss (1991d, p. 238):

> Clearly much more land, representing a larger array of ecosystem types, should be protected as wilderness. Existing wilderness boundaries need to be enlarged, and unnatural disturbances such as livestock should be removed. Where legally definable wilderness is lacking, "wilderness recovery areas" can be created by closing roads, actively restoring habitats and natural disturbance regimes, and reintroducing extirpated species.

Thus, while many writers correctly note that scenic and recreational values have driven many wilderness designations, the *idea* of wilderness has included a recognition of the importance of biotic protection. Many have seen wilderness as "the last home ground of countless species that would otherwise be doomed" (Frome 1974) and have appreciated that its very wildness, its lack of human-introduced "violence" (Leopold 1942), enables it to have unique biological quality.

Although wilderness has been closely linked in argument and philosophy with biological integrity, it has rarely been so linked in practice. Despite the biological roots for the importance of wilderness chronicled by Leopold, the withdrawal of specific Wilderness Areas from commodity production led to selections "biased toward low-diversity lands such as alpine zones" (Noss 1991d) and created institutional resistance to expanding wilderness. The growth of timber harvesting volumes in post–World War II years allowed those in the Agency who preferred Pinchot's viewpoint to come to the fore, bringing with them a growing distaste for wilderness. As G. Robinson (1988; p. 29) tells us:

> In the 1950's, it quickly became apparent that through this procedure [hearings on the boundaries of primitive areas under Regulation U-1], the timber industry representatives were persuading the Forest Service to delineate the boundaries of primitive areas to exclude any lands of commercial value. Frequently, sizable areas of land without timber or other exploitable resources were added to primitive areas. In this way, although the acreage of wilderness increased, key forestlands became accessible to industry. Dave Brower, then executive director of the Sierra Club, called this "wilderness on the rocks."

Senator Hubert Humphrey introduced the first bill to legislatively recognize wilderness in 1956, but it was opposed by then–chief of the Forest Service, Richard E. McArdle, who raised concerns about wilderness being in conflict with multiple use (Wilkinson and Anderson 1985). The Agency responded to Humphrey with proposed multiple-use, sustained-yield language, which eventually served as the core of the Multiple-Use Sustained-Yield Act of 1960 (MUSY). The Sierra Club did not oppose MUSY (Steen 1976) because of the inclusion of language stating that wilderness was consistent with multiple use (16 U.S.C. § 529).

By 1961, the Agency no longer opposed legislation that would legislatively recognize wilderness, and in 1964 the Wilderness Act (16 U.S.C. §§ 1131 et seq.) was adopted. However the process of evaluating areas for inclusion in the Wilderness System, carried out through the Roadless Area Review and Evaluation (RARE I) completed in 1972 and RARE II completed in 1979, and the litigation over the competency and biases of the Agency's work in these evaluations, led to an ever more contentious atmosphere regarding wilderness. Because of the increased scrutiny and growing concerns among the general public about Forest Service management that surfaced in the 1960s and early 1970s, wilderness became viewed by some as a way to protect lands from the intensive logging regimes that had arisen in the postwar period and thus represented in fact and perception an overt rejection of Forest Service stewardship. Supreme Court Justice William O. Douglas lauded wilderness in his writings and felt that multiple use was a "disguise for mercenary intentions" (Fox 1981).

The National Forest Management Act (NFMA) included wilderness as a specific component that must be included in forest plans [16 U.S.C. § 1604(e)(1)]. However, in practice, the Forest Service skirted the need to integrate wilderness within the planning goals in the NFMA process by negoti-ating many state-by-state wilderness bills prior to the completion of Forest Plans under NFMA. In 1984, Wilderness Areas were established in 20 states (including Wisconsin) out of the roadless areas addressed in RARE II. In Wisconsin, the Agency bargained vigorously to limit the size of Wilderness Areas and reluctantly supported the ultimate compromise bill in 1984 (Wisconsin Wilderness Act of 1984, P.L. 98-321, June 19, 1984). The Nicolet Forest Plan issued in 1986 proposed some of the most intensive log-ging levels in the entire forest to be carried out in the small Semi-Primitive, Non-Motorized areas in order to ensure that these areas would not be eligible for wilderness consideration in the next NFMA planning round.

The Agency's growing resistance to "withdrawal" of lands from multiple use through "lock up" in Wilderness Areas in the postwar years has created significant impediments to the argument that wild lands could serve as an im-portant component in protecting biodiversity. Rather than seeing conserva-tion biology as consistent with Leopold's concern for ecological "health," and acknowledging Leopold's roots in the Agency as well as the Agency's early sponsorship of the wilderness idea, we have found the Agency actively using the notion of wilderness as a shield against any further duty to preserve large contiguous blocks of habitat for the sake of protecting biodiversity. Since Wilderness Areas had already been negotiated in Wisconsin, the need for ad-ditional undisturbed areas was rejected as merely an attempt to reopen the Wilderness Area selection process rather than considered on its biological merits. And by confining concerns about diversity protection within the boundaries of designated Wilderness Areas, the Agency and its commodity customers felt free of any "constraints" on timber harvesting arising from bio-diversity or other wilderness values when addressing uses for the remaining lands in the forests. Other commentators have recognized the phenomenon that creation of "set asides" engenders a feeling of free license elsewhere (Brussard et al. 1992).

How then might the idea and practice of wilderness protection fit with our efforts to define a new diversity policy? First, it is important to recognize that from a biological viewpoint, Congressionally designated Wilderness Areas and the undisturbed areas needed to protect biodiversity do not fully overlap in purpose and should not be considered congruent. The quantity, shape, and location of land needed to serve the needs of biological diversity may not co-incide with the varied measures applied in selecting specific Wilderness Areas. And selection for biotic criteria may differ from traditional notions of "quality" wilderness driven by aesthetic or recreational attributes—either of which might well be of no significance in a well-chosen Diversity Maintenance Area. Indeed, even the definition of wilderness in the 1964

Wilderness Act treats ecological value as of secondary importance and of only scientific and educational value rather than a central purpose [16 U.S.C. § 1131(b)]. A new diversity policy can and must be justified with its own independent biological rationale and criteria.

Second, one must also carefully distinguish between the abstract *idea* of wilderness and its associated values as described by authors such as Nash and Oelschlaeger and the *reality* of Wilderness Areas as the product of a contentious legal and political process. Rather than harking back to the vision of Leopold, Carhart, or Marshall, the mention of wilderness is more likely to trigger painful memories of RARE I and II and public debates over particular wilderness bills. The commodity pressures of the postwar era, coupled with a new resourcism, have obscured the Agency's earlier role in recognizing the value of wilderness. Indeed, in the Wisconsin litigation, the Agency has taken the position that multiple use and natural (wild) ecological conditions are, by definition, at odds with one another (see Chapter 13).

With these important distinctions between wild conditions and Wilderness Areas, we should nonetheless recognize that a new diversity policy should be grounded in significant part on the idea of wilderness. Such a policy must be strongly tied to Leopold's appreciation of the role of wild nature as a critical element in protecting communities. A new diversity policy should be cognizant of the fact that this country had the political will to designate Wilderness Areas, albeit reluctantly, largely on Forest Service land. Moving forward from these bases, a new diversity policy should redefine the public's and the Agency's consciousness of wild conditions as a biological imperative which transcends the legal, political, recreational, and aesthetic senses in which the notion of Wilderness Areas presently exists in our culture. This new consciousness of wild conditions, coupled with the very low cost of such a management method, may build upon and revise the earlier Wilderness Area designations established largely for other purposes, and provide a key component of a new diversity policy.

The ancient forest debate: Is this a source of new policy for diversity?

The Pacific Northwest old-growth controversy is a major force in developing modern forest management and biodiversity policy for a series of reasons. The Agency's recalcitrance (as well as that of the Bureau of Land Management) in developing plans for recovery of the northern spotted owl and its willingness to invoke the economic exemption from absolute protection of a listed species permitted by Section 7(h) of the Endangered Species Act have helped pierce the multiple-use myth and heightened public awareness about species loss and the federal government's relativism toward endangered species.

More positively, the owl controversy represents a precedent for the use of the principles of conservation biology in conservation planning. For the first time, policymakers were presented with options developed in the first instance on the basis of reserve design theory and graded for their relative likelihood of protecting ecological values and "old-growth dependent species." Some of the scientific literature generated by the controversy has had a profound impact on the understanding of the adverse impacts of fragmentation (Harris 1984; Franklin 1989; Maser 1989; Moldenke and Lattin 1990a,b) and has given increased visibility to the importance of basing forest management on sound scientific information. While many bemoan the fact that scientific recommendations have been modified by political concerns, the role of scientific analysis has been strikingly more important in the Pacific Northwest old-growth controversy than in the typical forest management decisional process.

Despite these critical contributions to the debate over temperate forest management and species protection in the National Forests, there also are limitations on the value of the Pacific Northwest old-growth controversy in developing the foundations of a sound diversity policy. The issue in the Pacific Northwest forests is primarily one of protecting remnants of visually stunning, pristine lands still in near presettlement condition. In most other areas of the country, it is unlikely that national press coverage, presidential involvement, or the funding of intensive scientific studies would guide the fate of biodiversity.

A second limitation on the utility of the Pacific Northwest as a model for biodiversity policy is its grounding in the concern for attractive, individual species. Although many scientists and environmentalists have recognized that the viability of entire ecosystems is at stake, the most effective weapon for saving ecosystems in the short run has been the familiar paradigm of single species protection, either under the Endangered Species Act or the population viability standards of the wildlife regulations governing the Forest Service. This shortcoming of the owl dispute was particularly true during the 1980s, when press accounts continued to cast the problem as simply owls vs. jobs. The controversy has evolved as other bird and fish species become listed as threatened or endangered; but in the minds of many, the tangible cost of failing to protect remaining old growth is tied to one or two listed species, rather than to the old-growth community itself.

A third difficulty in drawing insight from the Pacific Northwest controversy is that it is so heavily laden with regional, political, and cultural pressures. For years the Forest Service, politicians, and industry ignored clear signs of ecological damage. This fostered continuing investments both personally and commercially in unsustainable management plans, and the pent-up economic, cultural, and political pressures now greatly limit and warp policy options. While any diversity policy must ultimately be tested against the full range of political concerns, we should not construct the first principles of a new policy on a foundation of political compromises.

In April of 1993, President Clinton convened a "Timber Summit" in Portland, Oregon. In early July that same year, the Clinton Administration announced a compromise package designed to give timber, environmentalists, and the Congressional delegation from the region just enough incentives to sign on rather than risk further litigation or legislation that might deprive one side or another of even such modest benefits (Egan 1993). The July 1993 Draft Supplemental Environmental Impact Statement (DSEIS), prepared jointly by the Forest Service and the Bureau of Land Management, proclaims an "ecosystem approach" to management of late-successional and old-growth forests in the Pacific Northwest (USDA Forest Service and BLM 1993).

The preferred alternative in the DSEIS, Alternative 9, contains new types of land allocations, such as Late Successional Reserves and Adaptive Management Areas, which allow timber harvesting to "accelerate the development of old-growth habitat characteristics" and to "learn more about ecosystem management," respectively. The "reserves" are not equivalent to Diversity Maintenance Areas since salvage and thinning operations would be permitted, even though such actions have been utilized in the past by the Forest Service to reap substantial volumes of timber. Even if harvested in smaller patches than traditionally used in the region, these logging entries could well lead to substantial fragmentation of the remaining old growth.

In sum, the Pacific Northwest old-growth controversy is encouraging to our efforts to define a new diversity policy because conservation science has been greatly elevated in importance in setting management goals, a factor we will conclude is central to a new diversity policy (Chapter 14). On the other hand, the tremendous political pressure in the region has encouraged reliance on highly experimental efforts to combine continued logging and restoration of late successional conditions. The notion that we can have it all from all acreage, sooner or later, through more and more engineering rather than more and more humility at the complexity and content of ecosystems, is a modern version of a familiar pattern in American forest management. Even with millions of hectares of pristine old growth in which to map Diversity Maintenance Areas, management alternatives that are based on the speculative benefits of newer, better engineering (albeit based largely on conservation biology rather than just silviculture) seem to remain more attractive to managers and politicians than the preservation or restoration of wild conditions.

Is There Support for a New Biodiversity Policy in the Existing Statutory Framework That Governs the Forest Service?

Any new diversity policy needs to be developed within the context of the three major statutes adopted in the postwar period for the protection of forest species. We will examine these three statutory schemes and determine what they lend to the development of a new, affirmative diversity policy.

The National Environmental Policy Act: Will this be the extent of our national environmental policy, and can it provide real protection for biodiversity?

The National Environmental Policy Act of 1969 (NEPA), signed by President Nixon on January 1, 1970 (42 U.S.C. §§ 4321 et seq.), requires that major federal actions significantly affecting the human environment be subjected to close scrutiny. Actions such as Forest Plans must be accompanied by a public disclosure document, called an Environmental Impact Statement (EIS). In theory, the process of identifying and analyzing the environmental consequences of a range of alternative actions will cause federal agencies to make decisions that are more protective of the environment. Disclosure of adverse impacts may heighten public interest in a project, giving agencies more input on concerns for environmental protection. An Agency's response to the NEPA process may be to choose an alternative that has fewer adverse impacts or, more commonly, to add mitigation measures to the alternative that the Agency preferred at the outset of the NEPA process.

NEPA has served two major functions in the decision process. For some time after its passage, NEPA was not well understood or internalized in Agency procedures. Many projects were blocked simply because the Agency failed to complete the necessary documentation of its decisions or failed to comply with some other major procedural requirement. In many cases, the procedural delays resulting from a NEPA challenge allowed publicity about the project, a more complete analysis of its effects, or simple politics to catch up with the Agency decision-making process, ultimately halting the project for reasons other than noncompliance with NEPA. A classic case of this type was the success of the Sierra Club Legal Defense Fund in halting a ski resort and related development in the Mineral King Valley (Turner 1990).

Second, NEPA does have the potential to affect substantive decisions. This may simply amount to more extensive mitigation plans, or it is possible that a series of NEPA compliance exercises may lead an agency to reevaluate its overall policy in an area. This apparently occurred in the case of the Forest Service's Region 6 herbicide policy after a series of five legal actions (O'Brien 1990). And when he was fired as Region 1 Regional Forester, John Mumma cited the environmental concerns disclosed by the NEPA analytical process as tangible factors which had limited his ability to accomplish the timber outputs sought by his superiors (High Country News 1991).

Although NEPA has influenced the outcome of a number of projects, and was intended to be "action forcing," many Agency staff have openly stated that they view the law's requirements as merely procedural, for the benefit of public involvement, but not necessarily imposing any substantive burden on the Agency to choose alternatives that are more beneficial from the standpoint of environmental protection. It is likely that this view is an outgrowth of guidance given to Agency personnel from their legal staff, who are familiar with the miserly reading that the law has received from the U.S. Supreme Court (Sutherland and Beers 1991).

For example, in its most emphatic limitation of the reach of NEPA, the Court in *Robertson v. Methow Valley Citizens Council*, 490 U.S. 332 (1989), held that NEPA did not require environmentally protective decisions, only informed ones. In considering the possibility of losses of mule deer in connection with the development of a ski resort, the Court made it clear that any decision regarding loss of deer would be acceptable under the Court's view of our national environmental policy provided the loss had been adequately described and considered:

> [F]or example, it would not have violated NEPA if the Forest Service, after complying with the Act's procedural prerequisites, had decided that the benefits to be derived from downhill skiing at Sandy Butte justified issuance of a special use permit, notwithstanding the loss of 15 percent, 50 percent, or even 100 percent of the mule deer herd. (*Id.* at 351)

With this view of the law, it is not surprising that many Agency staff feel that if they simply go through the motions of describing environmental impacts in some reasonable way, they remain free to make any decision they wish, regardless of its environmental impact.

This reaction to NEPA has been evident in the specific context of the biodiversity issue. In *Marble Mountain Audubon Society v. Rice*, 914 F.2d 179 (9th Cir. 1990), the plaintiff environmental groups argued that the Forest Service had failed to address the biogeographic implications of logging in a corridor between two remote, unfragmented areas. When the Ninth Circuit affirmed that the Forest Service had failed to take a "hard look" at these environmental impacts, the Agency's chief ecologist shrugged off the decision by saying the Agency's conduct was found lacking by the court "on the basis of a procedural matter" (Salwasser 1992).

In our view, NEPA's requirements for analysis and disclosure are not as "merely procedural" as its critics often assert. NEPA seems particularly well suited to promoting the application of conservation biology to the analysis required by an EIS in support of a National Forest Plan. The regulations promulgated by the Council on Environmental Quality (CEQ) under NEPA call for "accurate scientific analysis" [40 C.F.R. § 1500.1(b)] and "scientific integrity" (40 C.F.R. § 1502.24) in the analytical process. And in the event an agency felt it could not obtain needed information to predict foreseeable impacts, the regulations (40 C.F.R. § 1502.22) direct agencies to disclose to the public and agency decisionmakers "existing credible scientific evidence" and the agency's evaluation of adverse environmental impacts "based upon theoretical approaches or research methods generally accepted in the scientific community."

Reliance upon current evidence and generally accepted scientific theoretical approaches is particularly important in the biodiversity debate because many of the environmental effects predicted by conservation biology are probabilistic or depend on longer-term processes. Studies to provide definitive proof of some of the adverse effects of fragmentation could take decades or longer. In this situation, as with the problem of global warming, it is critical that sound scientific predictions be valued and not discounted as insufficient proof of environmental impacts until the damage is irreparably along its course. NEPA not only elevates scientific opinion to a position of great importance in the analysis but, by specifically addressing the situation in which environmental impacts are difficult to predict, acknowledges the importance of using the best scientific evidence available to predict likely future consequences.

A second value of NEPA, in contributing to a diversity policy, is its emphasis on an ecosystem rather than simply a species-by-species approach. The basic purposes of NEPA include enrichment of "the understanding of the ecological systems . . . important to the Nation" (42 U.S.C. § 4321). NEPA's declaration of national environmental policy recognizes the "profound impact of man's activity on the *interrelations of all components of the natural environment,* particularly the profound influences of . . . resource exploitation . . ." and the "critical importance of *restoring* and maintaining environmental quality . . ." (42 U.S.C. § 4331; emphasis added). Unfortunately these introductory statements of purpose have not yet been accorded significant weight by the courts.

However, language in the CEQ regulations may be more difficult for the courts to ignore, particularly if presented in the context of environmental impacts stated specifically in ecological terms. These regulations drafted in 1978 were surprisingly prescient in recognizing the three components of current diversity analysis—composition, structure, and function—when defining the environmental effects that needed to be analyzed and disclosed in an environmental impact statement:

> Effects and impacts as used in these regulations are synonymous. *Effects includes ecological* (such as the *effects* on natural resources and *on the components, structures, and functioning of affected ecosystems*), aesthetic, historic, cultural, economic, social, or health, whether direct, indirect, *or cumulative.* (40 C.F.R. § 1508.8; emphasis added)

When these references to ecology are coupled with NEPA's familiar requirements that agencies take a "hard look" at cumulative impacts of a pattern of agency activities, NEPA seems tailor-made for mandating the use of conservation biology, and an ecological approach in general, in assessing losses in biodiversity resulting from federal action.

Some have favored more explicit language in NEPA specifically requiring that environmental impacts be defined to expressly include effects on biodiversity. We contend, however, that NEPA's broad framework already contains several points of reference for the argument that a competent Environmental Impact Statement must now incorporate the best available "credible scientific evidence" and examine the cumulative impacts of forest fragmentation.

The species-by-species approach of the Endangered Species Act is being overtaken by the recognition of more pervasive biological forces

The Endangered Species Act (ESA) ranks with the designation of Wilderness Areas as the two most effective statutory means to protect biodiversity in this country. Although chronically underfunded and failing to save a number of species targeted for recovery efforts, there have been notable successes. But while the value of the Act has been and continues to be substantial as our only line of defense to continued extinctions, our present inquiry is whether the statutory framework of the Act should serve as a principal component of a broader diversity policy. In this endeavor, the logical first question is whether the Act might suffice in and of itself as a program to protect diversity, assuming that its funding and staffing problems were solved. Regrettably, the answer is no.

As Congress approaches reauthorization of the ESA, a great deal has been written about the Act's shortcomings along with suggestions for its improvement (Kohm 1991; Winckler 1992). We will address suggestions for improvements in Chapter 14, but first we need to examine the Act in its present form and interpretation. The first problem with the Act is that its listing process acts as a restrictive gatekeeper to whatever protections the Act may offer. The backlog of species awaiting listing is disgracefully large, leaving scores of species unprotected and unconsidered in environmental impact analyses. Although this problem might be addressed through increasing funding, or by adding a default provision that would waive a species onto the threatened list automatically if no action had been timely taken on its candidacy, there remain other problems with the Act's approach to diversity protection.

As implemented, the Act has not employed any mechanisms to assess which species are likely to become listed in the future and thereby triggering programs to prevent those likely future listings. There is brief mention of concern for species of neotropical migratory birds that may become endangered or threatened, in the language of the Act incorporating the Western Convention [§ 8A(e)], and Section 5 calls for the establishment of a "program to conserve fish, wildlife, and plants, including those which are listed . . ." (a provision that has not been implemented to date). But under the operative provisions of the Act, species moving toward threatened status are not addressed, and ecosystems are not afforded protection as such. It may be that the species most likely to be listed in the future are in the same habitats as species that are already listed; however, this is by no means certain. And to the degree that there is overlap in the habitats of listed species and those that are moving toward threatened status, it is not at all clear that protective measures that would address the already listed species would help prevent other species in the same ecosystem from moving toward threatened status.

We increase the risk of losing more species if our protective schemes take no account of regeneration problems until species are known to exist in the first place, and then recognized to be in dire difficulty. By "waiting until all the data are in," populations already are reduced and the forces of habitat loss that threaten them may already be locked into landscape patterns and uses for quite a long time. This is likely the case with species threatened more by pervasive fragmentation than specific destructive actions (e.g., logging through the specific patch harboring a rare plant). Restoring a meaningful unfragmented landscape surrounding the site of a threatened species population may simply take too long to permit recovery.

The next hurdle to biodiversity protection under the Act is the fact that it does not address the viability of habitats themselves, but rather the viability of individual species. Although Section 2 of the Act acknowledges that the purposes of the Act include "a means whereby the ecosystems upon which endangered species and threatened species depend may be conserved," the operative language contains no direct protection of ecosystems.

The Act indirectly protects habitats in a number of ways. Habitat may be designated as critical to a listed species [§ 4(b)], it may be protected under the terms of a recovery plan for a particular species [§ 4(f)], a mitigation plan adopted as a "reasonable and prudent alternative" to proposed government action [§ 7(b)(3)], or as part of a habitat conservation plan adopted to qualify a private party for a permit to commit an incidental taking [§ 10(a)]. However, these various means for protecting habitat have in common the fact that they are all directed at the needs of the threatened or endangered species which is driving the applicable statutory process.

For example, under the recovery plan language in Section 4(f), as amended in 1988, plans are for the conservation and survival of the listed species, and do not address the viability of the necessary habitat or take measures to protect that habitat in its own ecological terms. Under Section 4(f)(1)(B), the secretary is to incorporate into each plan

> a description of such site-specific management actions as may be necessary to achieve the plan's goal for the conservation and survival *of the species.* (Emphasis added)

Thus, the plan need only address and protect those features of the habitat which are thought to be necessary for the listed species. There is no provision that habitat be evaluated for its sustainability under the principles of conservation biology to determine its viability as a habitat in ecological terms. This then allows the ecological condition of the habitat to go unaddressed as long as those features of it presently thought to be important to a listed species are seemingly intact.

Moreover, the single-species approach to evaluating habitat may even call for management measures that are in the interest of the listed species but which *may run counter to the sustainability of the overall structure and function of the ecosystem* in which the habitat of the listed species is embedded. A concrete example of this problem is the recovery plan for the eastern timber wolf (USFWS 1978), complemented by a Wisconsin Timber Wolf Recovery Plan (DNR 1989). These plans recognize the wolf's sensitivity to high road densities and the human pressure that this brings to the forest. On the other hand, the majority of the wolf's diet in Wisconsin is thought to be white-tailed deer. In negotiations with the Nicolet National Forest over its wolf management prescriptions, both the Forest Service and the Fish and Wildlife Service insisted that they be able to continue clearcutting in wolf habitat in order to locally boost deer populations so that wolves had an ample supply of food.

While acknowledging that wolves were not prey-limited in Wisconsin (i.e., that other factors such as human pressure, population levels below minimum viable levels, and canine parvovirus were limiting), it was felt important to keep fragmenting the forest in the name of an endangered species. In the wolf recovery debate, no mention was made of the impacts of small clearcuts of aspen or of the density of the wolf-threatening road system on the overall integrity of the ecosystem.

Moreover, the species-by-species focus of the Act tends to play into the hands of its opponents since it suggests that the only thing at stake in a given dispute over listing or a recovery plan is a single species. In those frequent situations where listed species are in habitats that are more generally threatened, a clear public perception of the risks from the planned development activity might influence the level of acceptable risk or loss to biodiversity.

When the listed species is popular, such as the case with the eagle, whooping crane, or wolf, proponents of protection may take advantage from the charismatic nature of the species fronting for the ecosystem. When the species is a diminutive snail darter, American burying beetle, or homely Furbish's lousewort, opponents of protection ridicule the species and, by implication, the purposes of the Act. Having single species bear the weight of protecting entire ecosystems fails to either consistently muster public support or educate the public on the importance of protecting biodiversity. Perhaps media accounts of the broader biodiversity problem resulting from the 1992 Earth Summit or, more locally, the Wisconsin biodiversity cases will lend public support to a new diversity policy.

In conclusion, a comparison of the ecosystem approach of the preceding chapters with the species-by-species approach of the Endangered Species Act leads us to the conclusion that we need to move beyond the species-by-species paradigm embodied in the current text of the ESA and its past implementation. While it is important to retain and strengthen the Act as a stopgap measure, we must also move to develop and implement a broader diversity policy, whether lodged in amendments to the Act, new legislation, or a new policy which would affect the interpretation and implementation of current law.

Does the explicit recognition of diversity as part of multiple use in the National Forest Management Act (NFMA) represent the beginning of a meaningful diversity policy?

In discussing multiple use, we properly begin with the Multiple-Use Sustained-Yield Act of 1960 (MUSY). However, the unbridled discretion that MUSY granted the Forest Service allowed the Agency to manage with its own vision of multiple use, with few standards other than consideration of all uses. NFMA provided far more direction to the Agency, and for the first time expressly included the problem of biodiversity. In this section, we will explore the development of the basic viewpoints of the Agency toward that explicit diversity mandate.

In the extensive and authoritative volume on the legislative history of NFMA by Wilkinson and Anderson (1985), there is precious little information to guide one in assessing Congressional intent behind the diversity language. The Agency has argued that the purpose of the diversity language was solely to bar the practice of wholesale conversion of naturally occurring tree species to monocultures. This was an explicit concern of Senator Bumpers of Arkansas; however, there were broader concerns for ecological integrity motivating certain key drafters. Both Senators Humphrey of Minnesota and Randolph of West Virginia expressed concern in their proposed amendments for natural forest ecosystems and for the elevation of wildlife and ecological values relative to timber. Beyond these general statements, there is little evidence of the intent driving the inclusion of the diversity language, largely due to the fact that Congressional staff meshed several bills into the final Senate version, and these were in turn combined with House language (Wilkinson and Anderson 1985).

While evidence of intent would be helpful, principles of legal interpretation emphasize the plain language of a statute as dispositive unless it is ambiguous. The language regarding diversity states that forest plans must

> provide for diversity of plant and animal communities based on the suitability and capability of the specific land area in order to meet overall multiple-use objectives, and within the multiple-use objectives of a land management plan adopted pursuant to this section, provide, where appropriate, to the degree practicable, for steps to be taken to preserve the diversity of tree species similar to that existing in the region controlled by the plan. [16 U.S.C. § 1604(g)(3)(B)]

This language clearly addresses both the problem of tree species conversion and the broader problem of diversity of plant and animal communities. Thus, the statute must be seen as reaching more than the conversion question, or the other language would be superfluous—a result not favored in the law.

When adopting this language, did Congress really mean "biodiversity" in the sense it was used at the 1992 Earth Summit? No doubt few if any members of Congress had any background in conservation biology when addressing the diversity language of NFMA. But as with ESA, their sentiments to save species were genuine and acted upon, even if in some ignorance of the measures that would actually be necessary in the field to conserve biodiversity.

In a highly unusual provision in NFMA, Congress indicated an unwillingness to fully entrust the matter of rule promulgation to the Forest Service, and thus provided that a Committee of Scientists be established to promulgate the first set of rules that would govern the application of NFMA [16 U.S.C. § 1604(h)(1)]. The Committee's definition of diversity adopted by the Agency (in 36 C.F.R. § 219.3) was not stated in terms of the diversity of tree species, or even just all species of plants and animals. It went further to recognize the idea of diversity of "communities":

> The distribution and abundance of different plant and animal *communities and species* within the area covered by a land and resource management plan. (Emphasis added)

Under this definition, to determine whether diversity is going to increase or decrease over time, one must assess not only expected population trends of individual species, but also expected trends in the viability of different kinds of communities, as ecological entities. Under such a definition, the Agency should not constantly replicate common habitats with high alpha (local) diversity if this means impairment and eventual loss of other types of diversity, thereby lowering the overall (gamma) diversity of communities.

Two sections of the NFMA regulations state how the Agency is to address diversity in the forest planning process. The first of these provisions deals primarily with the procedures whereby diversity would be integrated into the planning process. Section 219.26 of 36 C.F.R. states:

> *Diversity.* Forest planning shall provide for diversity of plant and animal communities and tree species consistent with the overall multiple-use objectives of the planning area. Such diversity shall be considered throughout the planning process. Inventories shall include quantitative data making possible the evaluation of diversity in terms of its prior and present condition. *For each planning alternative, the interdisciplinary team shall consider how diversity will be affected by various mixes of resource outputs and uses, including proposed management practices.* (Refer to § 219.27(g).) (Emphasis added)

This regulation maintains the statutory dichotomy of diversity of communities and tree species, and requires that diversity (defined as gamma diversity in Section 219.3) must be considered throughout the planning process and its fate must be considered under each of the management alternatives in the Final Environmental Impact Statement (FEIS).

The second diversity regulation addresses the substantive question of the manner in which diversity would actually be provided for. Section 36 C.F.R. 219.27(g) mandates preservation and enhancement of diversity and indicates that the diversity of the "natural forest" (i.e., the variety of functional, native ecosystems) was to be sought, to the extent possible, in the decisional process:

> *Diversity.* Management prescriptions, where appropriate and to the extent practicable, shall preserve and enhance the diversity of plant and animal communities, including endemic and desirable natural- ized plant and animal species, *so that it is at least as great as that which would be expected in a natural forest* and the diversity of tree species similar to that existing in the planning area. Reductions in diversity of plant and animal communities and tree species from that which would be expected in a natural forest, or from that sim- ilar to the existing diversity in the planning area, may be prescribed only where needed to meet overall multiple-use objectives. Planned type conversion shall be justified by an analysis showing biological, economic, social, and environmental design consequences, and *the relation of such conversions to the process of natural change.* (Emphasis added)

The natural forest clause (and the natural change idea) is astonishing lan- guage to find in Forest Service regulations. Perhaps grounded in an early NFMA draft from Senator Randolph, it stands in stark contrast to the Pinchot notion that it was critical to move forests from their wild, old-growth, uncultivated state toward "scientific forestry" in a fully regulated, cultivated state. Here we find an explicit directive to the Agency to maintain forests in a "natural," wild condition, or move forest composition in that di- rection. We will see in Chapter 13 that the Agency strenuously denies any such implications for this language. Whether this notion of natural forest di- versity can survive both litigation and the Agency's efforts to change these regulations is a central issue for the development of our national biodiversity policy.

These regulations were adopted in 1979, were still in place in 1986, when the Wisconsin National Forest Plans were appealed, and remain in place as of this writing. They impose several requirements on the Forest Service with re- spect to diversity that go beyond the minimum levels of protection for min- imum viable populations of particular species in the separate wildlife regula- tions found in 36 C.F.R. § 219.19. Under these two regulations, the Agency must understand what diversity means in the sense it is defined in the regula- tions (i.e., gamma diversity, including the different habitats in the forest) and how that diversity is affected by various management practices. The Agency is given a distinct goal—the diversity of the natural forest. Reductions in diversity must be specifically justified as necessary in order to meet other objectives.

Just as the Agency felt MUSY had only amplified and confirmed the multiple-use concept, it argued NFMA explicitly reaffirmed the importance of MUSY and its multiple-use core as guiding the NFMA planning process. The Agency had enjoyed virtual immunity from judicial criticism under MUSY, and thus its interpretations of this language followed the pattern that would allow the Agency to retain maximum discretion in preparing management plans.

Early in the NFMA planning process, at a conference specifically devoted to the diversity question, then-Chief Max Peterson asserted that the diversity language of NFMA "did not require any specific level of diversity" in the National Forests. Peterson suggested that diversity might well be something that would fall out of the multiple-use analysis rather than be a driving force in that analysis (Cooley and Cooley 1984). The "no specific level" argument was a *non sequitur* since neither the statute nor the regulations could be read to call for any "specific level," and of course they could not do so since the degree to which particular forests could retain or regain native diversity would vary widely depending on a whole series of conditions involving past and future uses unique to each forest. Thus, "no specific level" became code words for letting the Agency off the hook from any substantive requirement.

Under such an approach, there was no need to set any real objectives for diversity. As long as diversity was somehow "considered," performance in this area could never be measured or criticized. This form of analysis, reducing nontimber values simply to matters to be considered but not necessarily provided for, maintained the tradition of the Forest Service of interpreting the multiple-use concept according to its own precepts. As Wilkinson and Anderson (1985, p. 30) succinctly state: "However the Act [MUSY] only required the agency to give equal 'consideration' to all resources, not to administer them equally" (footnotes omitted). Thus it was hardly surprising that the Agency's first reaction to the diversity language was to assure its personnel that diversity, too, could be dealt with by mere consideration within the black box of multiple use, without any measurable standards.

This position ignored the fact that the Committee of Scientists felt it important for the Forest Service to improve diversity, and to do so with the natural forest as the goal. The Committee defined diversity to include diversity of communities, as well as species, and gave the Forest Service a clear vector of the natural forest to work toward. These concepts were not recognized by the Agency in its early readings of the diversity mandate and indeed remained disputed throughout the Wisconsin diversity litigation.

Summary

This then was the jumping-off point for the Wisconsin National Forest cases. We hoped the Agency would acknowledge an obligation to utilize the principles of conservation biology and would revise its management plans to protect significant blocks of habitat. In that endeavor, we encountered the issues addressed in this chapter, which had significant impacts on the Agency's positions in those cases, as well as our own. We now turn to those practical experiences and explore how the issues addressed in this chapter influenced specific decisions and actions by the Forest Service in those administrative appeals and litigation.

13

Case History: The Wisconsin National Forests

The ever-changing policy rubrics (Integrated Resource Management, New Perspectives, or Ecosystem Management) and the sophistry of journal articles by senior Agency staff (Salwasser 1992) are, by the Forest Service's own admission, calculated to "position" the Agency for favorable public perception. To assess the Agency's diversity policy, we should therefore look instead to the Forest Plans, Environmental Impact Statements, administrative appeal decisions, and legal briefs defending those actions in which the Agency defines the extent of its legally binding commitments to specific resources, determines intensities of development and fragmentation, and interprets the reality of multiple use and species protection on the ground.

Since the mid-1980s planning efforts recounted here, the Forest Service has begun study of the problems of forest fragmentation, and the authors have been advised repeatedly that the Agency is making substantial progress in its appreciation of the problems addressed in this book. Are the Wisconsin case histories relevant only as a historical relic of the Agency's views before Ecosystem Management became the order of the day?

Our extensive experience with the Agency suggests quite the contrary. Despite repeated oral statements that the Agency was committed to protection of biodiversity, and frank admissions that the environmental analyses upon which current activities are based were grossly inadequate in ecological underpinning, the Forest Service has adamantly refused to rethink the management plans generated in the mid-1980s to take account of these concerns. Moreover, the Agency has exhibited a strong reluctance in the early 1990s to accept any explicit legal obligation to protect biodiversity. It has stated a formal legal position that the restoration of wild conditions would be fundamentally contrary to the Agency's multiple-use mandate, and it has proposed regulatory changes that could sharply limit the Agency's responsibilities toward the protection of biodiversity.

In this chapter, we will follow the course of the Agency's treatment of diversity from the mid-1980s planning effort, the administrative appeals and subsequent litigation over the Forest Plans, and further regulatory developments directly related to the issues raised in the Wisconsin cases.

The Context of National Forest Planning for the Wisconsin Forests

The Wisconsin National Forest planning effort of the early 1980s was strongly influenced by both the broad forces that had, up to that point, shaped the Agency as a whole (discussed in the preceding chapter) and several important regional factors.

Two views of land health

The Nicolet National Forest covers approximately 265,000 hectares in northeastern Wisconsin, and the Chequamegon National Forest consists of approximately 344,000 hectares in northwestern Wisconsin. Both forests have significant private inholdings yet also contain large blocks of land in nearly complete federal ownership exceeding 60,000 hectares in several rectangular configurations.

These lands were almost completely ravaged by clearcutting during the heavy logging at the turn of the century. The forests were acquired by the federal government through the purchase of burned and cutover lands, thought relatively worthless at the time, from the private timber companies that had owned them. During the 1930s, the Civilian Conservation Corps worked diligently to help replant the forests, and the effort to "bring the forests into production," as the Forest Service refers to this period, through the replanting of harvestable species, has continued since that time.

As a result of this history, both Agency personnel and the timber industry widely held the view, as the planning process began in the early 1980s, that the conservation practices of the Forest Service had brought the land back from a state of ruin. Since the general condition of the land seemed to have continually improved under the Agency's care, it was assumed that it would only continue to do so with further management grounded in the same principles. In oral argument in the Wisconsin diversity litigation, the Department of Justice explained to the judge that "the Forest Service had grown these forests back," implying not only the Agency's complete control of natural processes but also its unmitigated success in restoring not just trees, but forests.

To the Agency and many of its local constituents, vigorous young tree growth, as part of a planned harvest cycle, coupled with booming wildlife populations were seen as characteristics of the "healthy" forests existing in the beginning of the 1980s, on the eve of NFMA planning. In contrast, slow growing mature trees, snags, and rotting logs on the forest floor (and the real or imagined specters of pests, disease, and fire) were viewed as signs of a "decadent," wasteful, and "unhealthy" forest.

This Agency and timber industry conception of a healthy forest was something quite different from Leopold's idea of "land health." In his 1941 essay "Wilderness as a Land Laboratory," Leopold stated that wilderness was the condition in which, for "immensely long periods," the "component species were rarely lost" and that the "disappearance of plant and animal species without visible causes despite efforts to protect them" was a sign of sickness in the land. He bemoaned the fact that these ailments persisted because "our treatments for them are still prevailingly local." What was "still prevailingly local" in 1941 remained knowingly, grudgingly so, some 50 years later.

The principal tools of the prevailingly local approach: Timber stands and tree cover

These critical differences between what Leopold viewed as a healthy forest condition and the vegetation that the Agency's well-intentioned efforts had "grown back" were not cognizable to the Agency in part because of the very small geographic scale (timber stands) and the single measuring stick (tree cover) employed as the foundation for its planning efforts. At the outset of NFMA planning in the early 1980s, the major source of data available to the Forests was the set of stand maps and associated tabular data bases that had been accumulated over the years. The "return of these forests to production" was carried out through a tracking of the condition of individual stands. Stands were mapped and coded to reveal the age, size, and species of predominant tree cover. In the upper midwestern forests, where one does not find the large pure stands of the western forest, timber stands are often 2–10 hectares per stand.

From these stand maps, Agency staff could pinpoint the predominant species mix and age classes of the trees for each tiny stand throughout the two Forests. The staff also predicted the growth rates and volumes of timber at various points in the future. These stand data became the building blocks from which conclusions about available timber, now and in the future, were developed.

But the stand data were given a great deal more significance than simply timber volumes and types. The Forest Service viewed these timber stand data as the repository of a comprehensive characterization of the Forests. Thus, not only the variety of silvicultural treatments and designation of suitable timber lands depended on these maps, but the road system, wildlife opportunities, and ultimately even broad characterizations about biological diversity all flowed from the alternative ways these stands might be silviculturally manipulated in various planning alternatives.

Although distinctions are often drawn between the broad programmatic nature of forestwide planning as distinct from the site-specific decisions of individual timber sales, in fact the first-round forest planning in Wisconsin (and elsewhere, we believe from our review of other Forest Plans) was essentially based on a simple summing of the information from many tiny stands. The Agency did not consider the possibility that the cumulative character of the Forests might be quite different from the mere aggregation of the timber species and age-class data from individual timber stands.

The added pressure of the deer lobby

Another problem in the Wisconsin landscape familiar to Leopold was the ecological impact of white-tailed deer. Again in counterpoint to Leopold, the NFMA forest planners viewed the ability of the modern forest to produce huge quantities of hunting opportunities as a major sign of a healthy forest and one that greatly boosted the imputed "net present value" to the public of the planning alternatives which maintained or increased deer production.

White-tailed deer hunting represents a significant economic and cultural element of life in Wisconsin. Every fall, between 600,000 and 700,000 hunters enter the woods on the Saturday morning before Thanksgiving, taking up positions studied for weeks before, often on land hunted for decades within the same family or group. Boys are noticeably absent from school the week of Thanksgiving, and thousands of telephone messages lie unnoticed until the week following the annual fall hunt.

The Wisconsin Department of Natural Resources (DNR) received fees for deer hunting licenses of $18.61 million for the 1991 fall season, which was an increase from the levels of $12 to $13 million in prior years (Legislative Audit Bureau 1992). The combined gun- and bowhunting seasons killed 419,427 deer in the fall of 1991 (DNR 1992), with the prehunt deer population for the fall of 1992 estimated to be 1.25 million and somewhat lower in the fall of 1993.

The deer hunting culture has created significant incentives for game managers to manage specifically with a view toward boosting local deer populations. Many wildlife managers are avid deer hunters themselves. They are also aware of the financial impacts the deer tag fees and local expenditures have on agency budgets and local economies. The clout of hunting interests in game management policy in Wisconsin is legendary. A second popular species, ruffed grouse, also benefits from artificially created forest gaps (wildlife openings) and young aspen stands, thus adding additional weight to the pressure for artificial wildlife openings. For many wildlife managers, employing the species–habitat relationship between white-tailed deer and wildlife openings (edge habitats created as a result of those openings) to manipulate deer populations on a local level constitutes the foundation and preoccupation of much of wildlife management in Wisconsin.

The timber industry has long recognized the fortuitous symbiosis between logging and the "wildlife" (i.e., game) benefits sought by the deer lobby (Chapter 8). Together, these groups have concluded that "Wildlife loves managed forests," a popular bumper sticker distributed by the Society of American Foresters in Wisconsin, appropriately showing deer and grouse springing from a recently cut forest.

Thus, with the sure knowledge that intense management could create forests, that small stands were the appropriate scale for examination of the forest, and that their "publics" wanted substantial timber production and deer hunting opportunities, the Forest Service planning teams embarked on their assessment of the environmental impacts of the first round of Forest Plans under NFMA.

Diversity in the Nicolet and Chequamegon Plans And EISs: Simple Variety of Trees and Modest Site-Specific Constraints on a Species-by-Species Basis

The two critical documents in dictating the management of any National Forest are its Forest Plan and the Final Environmental Impact Statement (FEIS) intended to describe the environmental consequences of the Plan. The two Wisconsin Plans divided the forests into Management Areas of different basic types, with each type receiving a numbered designation. The Plans contained a set of management standards and guidelines for each of these types of areas. Thus, in any area receiving the type designation "1.1," for example, the District Ranger knew that he or she was to conduct activities with a particular mix and proportion of timber types, silvicultural methods, road densities, size and frequency of artificial wildlife openings, visual quality objectives, and so on.

The Agency refers to this mapping as a zoning of the forest, but that analogy is only partially accurate. Zoning merely sets the parameters of permissible development. In contrast, the Forest Plans set specific goals for the desired development of each of these areas in extraordinary detail, down to specific acreages expected to be producing particular tree species, and harvested by particular methods during the planning period.

Actual levels of forest development may differ from Plan goals for several reasons. The development activities that will actually take place is often restricted by budgeting of the Agency at less than fully requested levels. The Plan goal for timber, called the Allowable Sale Quantity (ASQ), may be reduced by budgeting to a lower "target" funded by Congress and allocated among the Regions and forests. But such reductions do not affect the intensities of timber management prescribed for a particular area, just the timing, give or take a few years, when that area is reached for harvest. Consistency between the Plan and specific timber contracts and other activities is required by law [16 U.S.C. § 1604(i)], and therefore an amendment of the Plan is required before significant deviation is allowed. In the seven years between the adoption of the Nicolet and Chequamegon Plans and this writing, no substantive amendments affecting timber outputs, road densities, or other features of forest fragmentation have been proposed or adopted by either Wisconsin forest.

Even if budget levels do not meet the ASQ level, the areal extent of fragmentation may still be just as significant biologically. District Rangers move around to the various corners of the Districts setting up projects, without any explicit guidelines to aggregate cut or uncut areas. As we saw in Part II, the fate of sensitive elements of biodiversity may be affected as much by the pattern of harvest and other landscape disturbances as it is by the specific level of the ASQ or the actual timber target that emerges from the budget process. Indeed, we have seen in the Nicolet National Forest, where many timber sales were unable to produce projected volumes due to disease, drought, and overly optimistic growth projections in the Plan, that this lower level of harvest volume actually compelled the District Rangers to enter more acres in search of timber volume, causing a more extensive pattern of disturbance.

While Forest Plans do not specify the site-specific location of individual timber sales, this is of little consequence to the overall impact of the Plan on fragmentation. Even if small acreages are moved about from one side of the road to the other, or there are small groupings of cuts near natural openings, the overall character of the fragmentation in the forest is not materially changed.

In sum, the Forest Plan firmly entrenches the intensity of human disturbance on the land and dictates whether the pattern of those activities is to be a coarse aggregation of logged and reserved areas or a fine-grained, crazy quilt of scattered disturbances throughout the landscape. The influence of the Plan is profound due to its detail, its legal force, its 50-year planning horizon, and the slow regeneration of mature forest conditions.

Species protections were overlaid after timber goals were established,
becoming mere constraints on forest development

Because the planning teams viewed the land through the lens of timber stands, and were called upon to produce specific projections of timber outputs, these goals became inherently dominant. There were no tangible goals for biodiversity and thus no way for such concerns reasonably to compete with goals for timber, game species (boosted by openings), and the road network to serve them. Once the general pattern and intensity of timber harvesting, wildlife openings, and roads were known, measures to protect species from the impacts of human development of the forests were never more than mitigation measures. Species protections did not drive the levels or patterns of development activities, but rather were seen as constraints on levels and patterns of forest development derived from other objectives (cf. Chapter 9).

Diversity was measured as variety of tree species and age classes of trees

Because timber stand data was the best information that the Agency had about the condition of the land, it became the backbone of the analysis not only of timber harvesting opportunities but also other environmental conditions and impacts, including biological diversity. Thus, the Agency defined favorable diversity conditions as the presence of a desirable variety of tree species and age classes of trees.

This notion of diversity, called "vegetative diversity" and often simply "diversity," was measured in the FEISs for the two Wisconsin National Forests through highly quantified information dealing with the mix of tree species and the relative percentages of trees in various age classes. Both the Nicolet and Chequamegon National Forests took the position that the greater the variety of trees and age classes in various stands around the Forests, the greater the "diversity" in the Forest.

For example, the Nicolet FEIS showed dramatic reductions in the predominant presettlement category of northern hemlock/hardwood from 61.5% of the presettlement forest to less than 1% of the forest at the time of the Plan. In place of hemlock/hardwood's major share of the forest, many other cultivated species showed increasing percentages over the decades. This was viewed as an increase in diversity. Similarly, the Nicolet identified four land type associations (LTAs) which related soil conditions to prevailing vegetation. LTAs were compared on three parameters—mixed hardwood, pine, and aspen. The LTA with the most evenly distributed mix in these three broad categories (Pence-Vilas, with 23% hardwoods, 29% pine, and 23% aspen) was denominated the most diverse.

Based on these simple changes in variety and age classes, the Agency asserted that the modern forests had greater diversity than the presettlement forests which had existed in those same areas. The Chequamegon FEIS stated (page I-12):

> The Forest is more diverse now than in the late 1800's and early 1900's. The need remains, however, to alter natural succession in some areas in order to maintain species and age diversity. Without man's intervention, the diversity of timber types and age classes will decrease and cause a decrease in the diversity of wildlife species.

Forest planners then linked these differences in the mix of timber types with the wildlife in the forest. For example, the Nicolet FEIS stated (page 4-68):

> **Wildlife.** The more diverse and spatially distributed the vegetation is, the more varied and dispersed the wildlife will be. . . .
>
> **Threatened and Endangered Species.** The list of endangered, threatened and sensitive species are as varied as their habitat requirements. *The more diverse the habitat provided, the better assurance there is that any single species will be provided for as well.* All alternatives provide for special needs of sensitive, threatened and endangered wildlife species through standards and guidelines and the vegetative diversity provided. (Emphasis added)

The viewpoint that diversity improved with increased manipulation of the landscape to provide a variety of age classes and species fit well with the planners' goals of extending timber harvests and corresponding management for high game populations across the Forests. The correlation of these goals was plain in the Nicolet EIS:

> Of all the vegetative types, the aspen type has the highest overall value to wildlife. It is a seral, short-lived species which produces high within-stand diversity. It is especially important to the most abundant game species—the white-tailed deer, ruffed grouse, and the snowshoe hare. . . . For wildlife, the more equal the amount and spacial [*sic*] distribution of aspen and age classes the better. (USDA Forest Service 1986b, FEIS p. 4-59)
>
> The mixed hardwoods timber component has a major influence on wildlife because it occupies such a large percentage of land in each alternative. The general value to wildlife of this group is lower than aspen, but if managed even aged [clearcut], its within-stand diversity and hence wildlife value is maximized. (USDA Forest Service 1986b, FEIS p. 4-63)

Analysis and protection of species were conducted on the basis of single-species and site-specific mitigation plans—a "prevailingly local" approach

The Plans and FEISs for the two Wisconsin National Forests contained specific mitigation measures to limit adverse impacts to threatened and endangered species, and several additional rare species identified by the state of Wisconsin as candidates for future listing. These species were analyzed one at a time, and the most significant direct threats to each were identified. This species-by-species approach biased mitigation measures to the "prevailingly local" viewpoint deplored by Leopold. Standards and guidelines for particular sensitive species were exclusively site-specific in nature, moving logging back a few hundred feet to avoid a nesting site or altering one or two specific conditions in the vicinity of the local population to avoid wiping it out. For example, in the Nicolet Plan, calypso orchid, showy lady's-slipper, and other plants grouped with them were to be protected by controlling beaver and their dams. When a project encountered one of these species, the project team would ask whether beaver dams were in the vicinity and might be controlled to help these species.

Meanwhile, the larger-scale biogeographic factors affecting the populations of these species were never considered. Had the Agency first identified the ecological integrity of various habitat types, particularly those presently rare but once well-represented in the forest, and examined regeneration problems of dominant species within those communities, a much more comprehensive list of species and ecological processes would have emerged, and it would have been plain that site-specific measures were only part of the cure.

Unfortunately, many interior forest species were not given any protection at all in this scheme. If the Agency could not find local causes for regeneration problems, a very ambiguous statement was made about the methods for protection of the species. For example, a District Ranger seeking guidance on how to reshape a timber sale to protect Braun's holly fern, foam flower, Blackburnian warbler, or solitary vireo—species noted by the authors as suffering from an overall lack of interior forest—would find only vague assurances in the Nicolet Plan that "general standards and guidelines" would be sufficiently protective.

The Agency's method of treating sensitive or listed species amounted to an aggregation of individual, localized mitigation plans which took no account of the cumulative effects on the biota of the overall pattern and intensity of landscape disturbance. By confining itself to this "prevailingly local" attitude about biological processes and impacts, the Agency led the public to think that local mitigation measures were adequate to protect all the sensitive species in the Forests.

The Forests' methods for species inventories, selection of indicator species,
and monitoring handicapped their assessment of diversity

The Wisconsin Forest Plans and FEISs certainly looked beyond trees to other species of the Forests. However, these efforts were hampered by limited inventory data, choice of common species as indicators of habitat conditions, and an overreliance on management indicator species (MISs) to predict very long-term processes. We focus on the Chequamegon National Forest as an example of how these difficulties affected the planning process.

The Forest staff had access to inventory information relevant to the diversity requirement of NFMA, some through cooperative efforts with the Wisconsin Department of Natural Resources, which included

- a vertebrate species list

- quantitative population data for selected sensitive and game vertebrate species

- a rare vascular plant inventory (originally generated in 1981 under contract with the Wisconsin Department of Natural Resources and revised by the forest ecologist)

- an inventory of potential Research Natural Areas (sites protected for their value as examples of natural communities)

- general vegetation information (timber type, age, etc.) through the vegetation management information system (VMIS).

While these data provided a start toward enabling the Forest to evaluate present and future forest diversity, these inventories fell short of the information managers needed to accurately predict the effects of forest management practices. No full inventory of the Forest's approximately 900 species of vascular plants existed; only commercial tree species and rare species considered sensitive at the state level were originally inventoried and monitored. Although some lichens appear to be good indicators of old growth (Thomson 1990), no information was available for nonvascular plants, lichens, or fungi. Like all other National Forests, no species list of the immensely diverse invertebrate groups existed on the Chequamegon. While a full species list would be impractical, the forest staff could reasonably have been expected to have compiled lists of aquatic insects, butterflies, and other groups with potential utility as indicators, especially inhabitants of restricted and/or strongly management-affected habitats. Finally, little information was available to address the effects of fragmentation, shifts in rotation age, type conversion, or other results of management practices with potentially deleterious consequences.

The methods for choosing indicator species suffered from a number of difficulties. All indicators chosen by the Chequamegon National Forest were vertebrate animals. Only one of these was a reptile or amphibian, which was later excluded from monitoring during Plan implementation. No plants or invertebrates were selected.

Habitats were lumped together and simplified in order to arrive at those communities deserving of an indicator species. Communities of different successional age were lumped together, obscuring the pronounced differences in species that exist among stands of young, mature, and old-growth forests. For example, even though deciduous trees dominate over half of the Forest acreage, all these stands were lumped into just two classes: "young/mature hardwoods (aspen)," with ruffed grouse as an indicator, or "old-growth hardwoods," utilizing the pileated woodpecker. Young, even-aged stands of red oak and red maple, uneven-aged stands of sugar maple, and mixed basswood and yellow birch forests with an ironwood subcanopy were all lumped into a single category despite representing radically different environments for many constituent species. Tracking populations of ruffed grouse and pileated woodpeckers could never adequately represent ecological changes occurring in these diverse communities.

A preponderance of the indicators chosen for the Chequamegon National Forest were species able to occupy a range of habitats (habitat generalists) and not particularly sensitive to forest management activities. It makes little sense to monitor the thirteen-lined ground squirrel, ubiquitous throughout the Forest in the many openings, in that such species actually benefit from most timber harvest activities. Similarly, white-tailed deer, ruffed grouse, and the common yellowthroat all thrive in cutover areas, making increases in their numbers primarily an indicator of accelerated disturbance. Lumping old-growth and mature forests with early successional stands and representing the lot with a single indicator that is insensitive to, or benefits from, cutting has the effect of masking the impacts of timber harvesting within the forest. For another category of potentially sensitive species, those federally listed as threatened or endangered, the Forest Service deferred all responsibility to the U.S. Fish and Wildlife Service and the "recovery plans" for these species.

As a result of these simplifications, a majority of the chosen MISs were not in danger of extirpation, nor did they indicate anything about the condition of forest species that were. In fact, many MISs could be expected to benefit from the increase in wildlife openings and other edge habitats under the preferred Plan alternative in each forest, thus leading the public to see the Plan as protective of all plant and animal communities.

Throughout our discussions with Agency staff, they often responded to our concerns by indicating that future monitoring efforts would determine whether different management policies and landscape patterns were warranted. But these monitoring plans were not established in a way that could provide a fair indication of the processes of ecological dysfunction we had identified.

Appendix B of the Chequamegon Plan provided brief statements of intended monitoring techniques for a number of species. For some species, the Chequamegon planned to monitor populations at least in part by monitoring available habitat. This was a circular approach, since the purpose of monitoring populations is in part to determine the amount of available habitat. For other species, actual population surveys were to be undertaken. Many of the birds chosen as indicators were to be monitored as part of the Fish and Wildlife Service annual breeding bird surveys. Heavy reliance on these surveys, however, overlooked the importance of examining reproductive success. With the increasing realization that forest fragmentation can lead to reproductive failure for birds via nest predation or parasitism by cowbirds (Brittingham and Temple 1983; Wilcove 1985), it might have been evident that simple bird counting would be an imperfect measure of habitat suitability. While estimating reproductive success is a far more intensive undertaking, reliance on current population levels will result in artificially sanguine results for species experiencing regeneration problems.

In addition, the Forests might have examined habitat availability distinct from trying to track MIS populations. For example, in the upper Midwest, the rare calypso orchid is found almost exclusively in swamps dominated by northern white cedar, often in older stands. Although studies of this orchid might show the population to be stable at present, there was evidence in the Chequamegon planning documents that cedar was reproducing poorly. Clearly, if calypso's habitats fail to regenerate, calypso will decline as well. Thus, if the Forest had monitored calypso populations and not the condition of their habitat, this plant's demise could be nearly assured before the Chequamegon detected any declines. Similar scenarios likely hold for other species, especially those with potential time lags between habitat degradation and noticeable impact on populations. Thus, although appropriate monitoring techniques were sometimes used, broader and more sophisticated approaches were clearly needed to move beyond the characterization of these forests in terms of tree species and age classes that dominated the planning effort.

Underlying these specific shortcomings in assessing sensitive species, however, was a fundamental rejection of the value of scientific prediction. The decision to eschew the predictions of conservation biology was particularly imprudent given that many environmental phenomena resulting from forest fragmentation can be expected to take years to become manifest, that the

Agency admittedly did not have detailed knowledge of the biota it was managing, and that the Agency intended to continue to impose considerable artificial disturbances on the landscape.

MIS analysis and monitoring can be useful components in a system of environmental assessment grounded in ecological understanding. In the case of the Wisconsin National Forest Plans, however, the monitoring program did not grow out of a foundation in ecological understanding, and therefore it was destined to miss the species critically affected by forest fragmentation. And by relying so heavily on future monitoring, the Agency avoided addressing the strong evidence already in hand describing a series of critical adverse environmental effects on the native diversity of these forests.

The use of the concept "old growth" in the context of the Great Lakes states forests led to further confusion and misrepresentation about the fate of biodiversity

Those familiar with the history of the Great Lakes forests might be surprised to find the repeated use of the concept of old growth in the Forest Plans in Wisconsin. The term was widely used in the Wisconsin Plans and in the language employed by the Agency in legal briefs filed with the federal court, but with very different meanings and in different contexts. For example, when describing the pre-European settlement condition of the Chequamegon National Forest, the Agency's brief told the court:

> Prior to settlement of northern Wisconsin, the majority of the land
> that is now the Chequamegon was covered by old-growth stands of
> a variety of tree species.

Here, the Agency plainly meant a functional, late successional, ecological condition, sustainable for centuries on a forestwide scale, a meaning in keeping with the general, scientific use of the term (cf. Davis 1993).

A mere eight pages later, however, the Agency told the court, as they had told the public in the Forest Plan:

> . . . Alternative G [the chosen alternative that became the Plan] had
> a stronger focus on the creation and preservation of old-growth,
> projecting an increase in that component to 40 percent of the Forest
> in the future.

In this quotation, the Agency used the term as defined in the FEIS—"a stand of trees older than normal rotation age for the type that provides important habitat conditions not normally found in younger stands." Under this much more inclusive definition, aging stands of early successional species (including aspen over 40 years) qualified as old growth, although they bore no similarity to the tree species predominant in the presettlement forest and could not legitimately be seen as contributing to the solution to the regional loss of mature forests. The "important habitat conditions" qualification meant denning or nesting cavities in the trees, without reference to whether the forest, not just a few selected characteristics of the trees, exhibited old-growth characteristics such as those found in the presettlement Wisconsin forests.

These two quotations were each correct, standing alone and based on the different meanings of "old growth" in each case. But since the differences in the meaning of old growth in each case were not clarified by the Agency, the two statements (and the widespread Agency use of the latter silvicultural definition) created the false impression with the public and among Agency staff that the environmental benefits of functional climax forests could be obtained through the management of a highly scattered array of tiny, medium-aged forest stands.

Indeed, the old-growth policy of the Nicolet held that the best way to maximize the old-growth value of stands that were left beyond normal rotation age was to surround them with differing habitat and to leave them in as scattered a pattern as possible. The Nicolet supplement on old-growth policy, incorporated by reference into the Nicolet Plan, stated (page 59B):

> Secondly, old-growth designated stands are not set-aside small wilderness areas. *They are stands to be managed for timber, but at a delayed rotation age.* (Emphasis added)

At page 59C, the supplement stated:

> Forests contain the greatest number of wildlife species when they consist of a diversity of wildlife habitats. Therefore, forests that contain old-growth habitats, mature, pole-sized, young growth and open habitats will contain the greatest number and variety of wildlife species.

The supplement also included a set of criteria for selecting old-growth stands (pages 59D–59E). The following criteria were included (page 59E):

> 3. *Location.* One of the most important selection criteria for old-growth is proper distribution. Wildlife, especially birds, are very mobil [*sic*] and *scattered stands of old-growth distribute those species using the old-growth habitat.* . . . For wildlife, the following areas are some of the more desirable: . . .
> (b) *Transition areas* between two or more timber or vegetation types.
> (c) Small stands *differing significantly from the surrounding* timber types . . .

(e) *Isolated* or inaccessible stands which are difficult to manage.

(f) *Near permanent wildlife openings . . .*

(h) Old-growth stands should be *distributed evenly within and between compartments as much as possible.* (Emphasis added)

Thus in the Nicolet, where old growth could be actively managed (i.e., outside of Wilderness Areas), it was thought best if these patches were near edges (b, c, and f), in tiny patches (c, d), and as widely and evenly scattered as possible (3, above, and h), and eventually logged. All these recommendations are directly at odds with the consensus of scientific thought on the subject (Chapters 4–7).

Tiny Research Natural Areas were to serve as the specimens of ecosystems to support rare species

The Nicolet Plan also relied heavily on tiny Research Natural Areas (RNAs) to "represent natural ecologic communities. Their purpose is to promote and protect natural diversity in all its forms" (Nicolet FEIS page 3–8). The RNAs were denominated "8.1" areas in the Agency mapping and land designation scheme, and the general management description of 8.1 areas showed the heavy biological burden these areas were meant to shoulder under the Plan:

This goal will emphasize the following:

(a) The preservation of unique ecosystems for scientific purposes.

(b) The protection of unique areas of biological significance.

This management area occurs mainly in small tracts (usually less than 100 acres [40 hectares]) wherever management information has revealed the uniqueness of the site. The vegetative condition may range from a[n] undisturbed ecosystem to a condition highly modified by past actions. . . . (Nicolet Plan page 142)

These areas were tiny, with 18 sites in the Nicolet averaging 73 hectares and totaling 1316 hectares forestwide, or less than 0.5% of the 265,000 hectares in the Nicolet at the time of the Plan. As we discussed in earlier chapters, such tiny plots could not possibly constitute, much less preserve, unique ecosystems.

The Appellants' Challenges to the Plans and FEISs and the "Second-Generation" Diversity Policy

The mid-1980s techniques for environmental analysis tell us much about the fundamental Agency outlook on ecological issues and their relative importance in the multiple-use framework. Equally interesting is the Agency's reaction when confronted with a barrage of conservation biology literature and opinion in the Wisconsin Forest Plan appeals. In response to the appeals, the Agency reformulated its diversity policy, shying away from such obviously limited analytical tools as tree cover. But at bottom, the fondness for single-species mitigation plans and a refusal to commit to the protection of undisturbed areas of any size remained at the core of the Agency's approach to diversity.

The appeals and litigation over the Wisconsin National Forest Plans

The Draft Plan and Draft EIS for the Nicolet were issued in November 1984, and the parallel documents for the Chequamegon were issued in late February 1985. The Draft planning documents made no reference to the rare plant surveys prepared by W. Alverson and colleagues. These botanists, Professors Iltis and Waller at the University of Wisconsin Botany Department, and local activists with the Sierra Club and Audubon Society were struck with the extensive road network and proliferation of harvesting and wildlife openings called for in both Draft Plans.

Within the next several months, this group of activists (later to become appellants in the administrative process and plaintiffs in the federal court litigation) wrote many letters to both National Forests asking for consideration of much less intensive alternatives and large areas of reduced fragmentation. The Wisconsin Audubon Council wrote

> Larger sections of undisturbed, roadless areas are needed for threatened and endangered species. We feel that none of the alternatives give enough consideration to sensitive biotic systems; major emphasis is on timber development, not on ecosystems. . . .

Another letter told the Agency

> large completely roadless blocks have a biological value that is not addressed in your report, and on which a monetary value cannot be fixed.

Several of the letters asked the Nicolet to keep 25% of the Forest (over 60,000 hectares) in primitive unroaded conditions. In answer to the Chequamegon Draft documents, the citizens prepared a "Conservationists' Alternative," calling for an aggregation of harvesting activities and a corresponding aggregation of reserved areas termed Diversity Maintenance Areas, or DMAs (Chapter 11), comprising 17% of the Chequamegon (approximately 57,000 hectares). The activists sought meetings with the Forest Supervisors for the two Forests to explain their concerns and to suggest alternative management concepts that might mitigate the threats to diversity from the continual fragmentation of the Forests. In the Nicolet, the activists encountered an old-school forester in Supervisor Jim Berlin, who viewed the proposals for large blocks of unfragmented forest as a cynical attempt to "lock up" more Wilderness Areas. Berlin's response to the scientific arguments submitted to him was simply that these issues had been fully disposed of in the 1984 Wisconsin Wilderness Bill.

In the Chequamegon National Forest, Supervisor Jack Wolter was more receptive to concerns expressed about biodiversity. He instructed his staff to see if they could move some of the planned logging activities to other areas in the Forest without sacrificing timber production. Upon discovering that the Chequamegon could aggregate cutting, at least for the first several years, without having to lower ASQ, Mr. Wolter decided to set up two areas in which no further logging, road construction, or permanent wildlife openings would take place during the period of the Plan. These new areas were mapped and management prescriptions were developed prohibiting logging and other disturbances, but allowing a range of activities such as hunting, ATV and motorboat use, snowmobiling, berry picking, firewood cutting, and other non-consumptive uses such as skiing, hiking, birdwatching, and the like.

The proposed Diversity Maintenance Areas were designed to include areas of the two Forests that were relatively less disturbed; contained rare plant sites, desirable forest types, and minimum inholdings; and, where possible, surrounded the relatively small Congressionally designated Wilderness Areas (Fig. 13-1). By eliminating direct human-caused disturbances in the DMAs, regeneration was predicted to improve for those species preferring interior forest conditions. In time, the structure and function of mature forest habitat would once again exist in these large sustainable blocks, to the greatest degree attainable in modern Wisconsin, with meaningful biological consequences. But in order to secure the Supervisor's agreement to these areas in the final Plan, the activists agreed to a number of compromises from desirable DMA design principles (Chapter 11).

These proposals for the Chequamegon were transmitted to the Regional Office for final review before issuance as a part of the final Forest Plan. Because of concerns over the precedent that would be set in mapping certain areas as off-limits to logging, the Regional Forester, Floyd J. "Butch" Marita, rejected the proposal from Supervisor Wolter. Marita did so without consulting any scientists within the Agency, as was apparent from the documentary record and as admitted in oral argument by legal counsel for the Agency. Document inspections taken in the subsequent appeal process revealed that the Agency staff was specifically looking for a way to deal with the issue which would not be vulnerable to appeal. Hence, Mr. Marita directed both the Nicolet and Chequamegon Plans issued in August of 1986 to contain similar language regarding the Agency's views of diversity. The Plans and FEISs for the two Forests assured the public that biological diversity would be improved as a

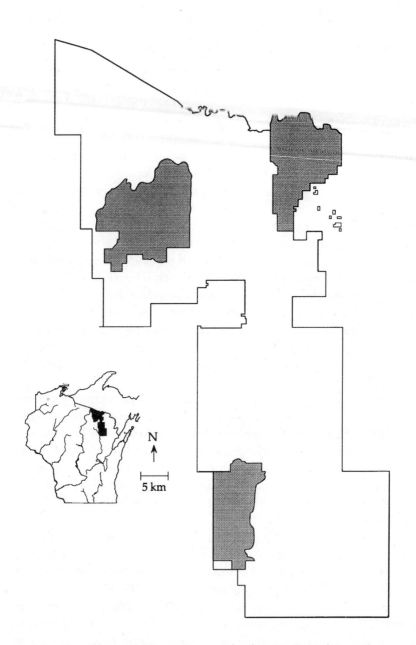

Fig. 13-1. Three Diversity Maintenance Areas of 16,000, 20,000, and 28,000 hectares were proposed to staff of the Nicolet National Forest in 1986 as specific examples of the way the Dominant-Use Zoning Model might be applied to long-term forest planning. None of the reserves were established, nor was there any comprehensive analysis of the potential effect of their establishment on the biological and economic goals in the Forest Plan.

result of the Forest Service management adopted in the Plans, but never rebutted the ecological impacts of concern to the activists.

In late 1986, the activists appealed the Wisconsin National Forest Plans to the Chief of the Forest Service, F. Dale Robertson. As appellants, they asked that the Chief find the Plans seriously lacking in biological foundation, and that he direct the Forests to redo the Plans and FEISs after taking account of the principles of conservation biology describing the environmental consequences (in long-range ecological terms) of alternative geographic patterns of forest use. He was also asked to direct the Forests to limit fragmentation by establishing Diversity Maintenance Areas, in which the management prescriptions would bar commercial logging, construction of new roads, reconstruction of minor roads, and the construction or maintenance of wildlife openings. Representatives of the timber industry, hunting organizations, and several local units of government voiced significant apprehensions about recognizing large reserved areas within the National Forests of Wisconsin and intervened in opposition to these appeals.

The appellants filed several hundred pages of record materials (WFCTF 1986) supporting their views that the Agency could not competently address the issue of biological diversity while failing to take account of major developments since the 1960s in conservation biology, including evidence from studies in population biology, community ecology, and biogeography. The filings included a long bibliography of the major literature in conservation biology, specific studies on nest predation, deer herbivory, edge effects in general, and other aspects of forest fragmentation. The appellants also filed letters of support from 13 nationally eminent scientists, including Jared Diamond, Paul Ehrlich, Daniel Janzen, Peter Raven, and E. O. Wilson, all of whom strongly supported the appellants' concerns about the adverse environmental effects of forest fragmentation. Many of these scientists specifically voiced support for the DMA concept and its rationale—that large protected areas were needed to allow regeneration of natural disturbance regimes and forest interior conditions.

Chief Robertson had an opportunity to state national policy on the role of conservation biology in forest planning when he decided the Nicolet and Chequamegon appeals. However, instead of affirming a place for conservation biology in forest planning, he specifically ruled that Agency staff had adequately accounted for material environmental effects notwithstanding the undisputed fact that they had not utilized many of the principles of conservation biology in their work. In essence, the Chief's ruling said that the methods of equating tree-cover (species and age-class) variety with diversity, providing tiny Research Natural Areas to protect sensitive habitats, and establishing an aggregation of mitigation plans for individual threatened and endangered species were sufficient to safeguard the biological diversity of the Forests.

One of the most astonishing aspects of the rulings in these cases was the fact that Chief Robertson never commented on the weight of scientific opinion from the experts assembled by the appellants. Just as Regional Forester Marita did not consult scientists for input on his decision to delete Supervisor Wolter's decision to include DMAs, the Chief concluded the Agency staff work was sufficient to address biodiversity without the benefit of any expert opinion supporting that view. Dr. Hal Salwasser, a frequent commentator on the subject of biodiversity for the Agency throughout the late 1980s and early 1990s, attended the oral arguments in the cases but remained mute on the merits. The forest planning record for the two Wisconsin National Forests reveals that several senior scientists in the Agency (Crow, Franklin, Ohman, and Salwasser) knew in the mid-1980s that forest fragmentation posed serious environmental impacts, yet the Chief's decision gave these materials no weight or mention. The Agency did not bring forth any expert, either within or outside the Agency, to rebut either the warnings from the conservation biology literature or the specific points from Drs. Diamond et al. This refusal to allow scientific materials and opinions to sway management choices was one of the most troubling aspects of the cases and has caused us to single out the elevation of science as a critical component of a new diversity policy (Chapter 14).

While the administrative appeals were pending, and for a short period after the Chief's decisions, the plaintiffs, the Forest Service, and a timber industry coalition agreed to convene formal mediation. The Agency proposed short-term reductions of fragmentation in certain areas but refused to adopt any basic ecological yardsticks for future management. Even these proposals were quickly withdrawn after timber industry objection, and the appellants decided that the Agency would not make material changes in their management unless compelled to do so by a court order. The appellants filed suits as plaintiffs in the U.S. District Court for the Eastern District of Wisconsin, before District Judge John W. Reynolds. The Nicolet case was filed in April of 1990, and the Chequamegon was filed in October of that year. When the cases went to federal court, the nature of available relief changed because the federal court could not order the implementation of specific management proposals. Therefore, in the lawsuits, the court was asked to rule on the fundamental legal duties of the Agency.

In their presentations to the court, the plaintiffs argued that significant developments in ecological science in the 20 years preceding the issuance of the Wisconsin Forest Plans had to be included in the analysis and public disclosure of environmental effects required by NEPA. The plaintiffs argued that the diversity language in NFMA and its regulations required that the Agency strive to return portions of the Forests to their "natural forest" condition, and that only by doing so would the overall diversity of the Forests be enhanced. Only when the products of artificial disturbance and the benefits of undisturbed interior forest communities were both provided could the Agency meet its responsibilities to provide diversity across the entirety of the land area governed by the Forest Plan. The plaintiffs argued that the Agency's formulation of its limited duty to address diversity in terms of individual species mitigation plans, tiny reserved Research Natural Areas, and future monitoring was not a legally sufficient response to the NFMA mandate to "provide for diversity of plant and animal communities."

The "second-generation" diversity policy for the Forest Service

When responding to the administrative appeals, and in legal briefs filed with the court, the Forest Service was forced to articulate its diversity policy and specifically to explain how that policy accounted for the cumulative impacts of the Agency's proposed management. While the issue of diversity was stated almost exclusively in terms of tree species and age-class variety in the planning work described in the Plans and FEISs, when forced to define a diversity policy in the face of the ecological challenge posed by the plaintiffs, the Chief avoided reference to the tree variety concept. The Chief shifted ground to a second-generation policy for treating diversity, which he pieced together from several aspects of the Agency's planning work that addressed wildlife and individual threatened or endangered species.

In deciding the Wisconsin Plan appeals, the Chief stated:

> Diversity establishes no additional requirements beyond those identified for management indicator species, for population viability of sensitive species, for the recovery of threatened or endangered species of plants and animals, and for protection of special habitats.

Although this policy abandoned the tree variety concept, it retained certain key aspects of the Agency's planning efforts in the first round of forest planning: (i) maintaining a species-by-species approach to sensitive species; (ii) protecting habitat for its own sake only when it represented a "rare or unique" example, such as a Research Natural Area; and (iii) promoting site-specific analysis rather than a large landscape perspective.

In addition to its continuing scientific and philosophical bias against a broad ecological perspective, this second-generation diversity policy also ran headlong into problems with the statutes governing the Agency. By confining the Agency's duties to tasks that arose under the Endangered Species Act, the discrete wildlife statutory language and regulations (the MIS and MVP concepts), or the Research Natural Areas regulation, the Agency was, in effect, trying to delete the separate diversity mandate in the statute and regulations. If the tasks the Chief outlined were sufficient to address diversity, why had Congress and the Committee of Scientists thought that there was a need for separate diversity language in the statute and regulations?

The plaintiffs emphasized that the separate diversity regulations had distinct and important requirements that went far beyond the Chief's narrow prescription for diversity. These included understanding the impacts of various planning alternatives on forestwide diversity (36 C.F.R. § 219.26) and striving to attain, to the extent practicable within the other demands on the land, the diversity "at least as great as that which would be expected in a natural forest" [36 C.F.R. § 219.27(g)]. As a result of not meeting the requirements of these distinct diversity regulations, not only did the Agency overlook key planning steps where ecology should have been accounted for, but every one of the Agency's representations about diversity, communities, wildlife habitat, old growth, and sensitive species (including consideration of habitat and populations under 36 C.F.R. § 219.19) was undermined because the Agency's work was not informed by conservation biology. And the standard for diversity protection adopted by the Chief ruled out meaningful analysis of cumulative or large-scale effects—the leading edge of contemporary conservation analysis.

The Agency has sought to conform the NFMA regulations to the Chief's narrow interpretation of the Agency's diversity responsibility

To ensure that the Chief's interpretation would control future disputes about the Agency's duty to provide for diversity, in early 1991 the Forest Service included far-reaching revisions of the diversity regulations in its proposal to rewrite the NFMA regulation, Advance Notice of Proposed Rulemaking (ANPR) (56 Fed. Reg. 6508; 2/15/91). As of this writing, this proposal has not been further promulgated as a draft or final rule change. Nonetheless, it reflects the Agency's clearest summary statement of its policy toward biodiversity.

The original regulatory requirements that forest management seek to provide diversity "at least as great as that which would be expected in a natural forest," and which allowed reductions in diversity "only where needed to meet overall multiple-use objectives" [36 C.F.R. § 219.27(g)] were to be scrapped under the proposal. Instead, the ANPR stated that the Agency could

> provide for diversity of plant and animal communities by identifying management indicators and specifying measurable conditions to:
>
> (1) Conserve threatened or endangered species . . . .
>
> (2) Maintain viable populations of native plant and animal species by designating in the forest plan sensitive species and providing standards and guidelines that will ensure their conservation when an activity or project is proposed that would affect their habitat.
>
> (3) Protect rare or unique biological communities.
>
> (4) Provide habitat capability needed to support populations of species at selected levels for commercial, recreational, scientific, subsistence, or aesthetic values.
>
> (5) Monitor trends in management indicators relative to goals, objectives, and standards and guidelines . . . .

The Nicolet and Chequamegon Plans and FEISs might well have qualified as meeting this proposed regulation. Points (1), (2), and (4) still permit and seem to even contemplate a species-by-species approach. Point (3) fits well with the RNA regulation and holds that measures to protect biodiversity are only appropriate once a type of habitat has reached the point of being rare or unique. This latter provision seems to parallel an unfortunate aspect of the current application of the ESA by waiting until a community is degraded to the point where ecological processes are degraded and species are being locally eliminated before protective action is triggered. NFMA does not seem to limit the application of the diversity mandate to such a late date.

In this ANPR proposal we also see the continued overconfidence evident in purporting to be able to manage particular species at "selected levels." The emphasis on monitoring should be a separate requirement that overarches all forest management and not be considered a substantive achievement to be credited toward saving biodiversity. The five proposals are prefaced by the idea that the Agency need not necessarily accomplish the proposals, but merely identify management indicators and specify measurable conditions that would address them. If the Agency reasonably picked indicators, could it be criticized for failing to actually conserve species? If the Agency could not specify a measurable condition for the gradual loss of biodiversity, could it be challenged over the gradual loss of regenerative capacity, or any other ecological dysfunction, of the tiny remaining patches?

It should also be noted that the ANPR proposed to codify the Agency's position that irreversible and irretrievable commitments of resources only occur at the project, not the forestwide level. This portion of the new rules would seek to make the "prevailingly local" viewpoint of biology the law of the land. If made a part of the Code of Federal Regulations, the Agency might argue in the future that large landscape ecological processes were immune from legal critique.

The plaintiffs filed extensive comments on the ANPR, fearing that even if their cases were ultimately successful in court, the Agency might use the ANPR to complete an end run around the legal duties established by the court. In those comments, the plaintiffs expressed particular concern that the limited duties the Agency was willing to accept under the ANPR would relieve the Agency of any obligation to restore the native biological heritage of more than a half million hectares of northern Wisconsin.

Are Multiple Use and Natural (Wild) Ecological Function Inalterably at Odds?

The last round of legal briefs in District Court in the Chequamegon case brought us full circle to the old standoff between the views of the land articulated by Pinchot and Leopold. When the plaintiffs challenged the Agency's narrow reading of the diversity regulations as excluding ecological considerations and specifically failing to account for the statement about achieving "natural forest" diversity, the Agency candidly articulated its true position. In its answering brief in the Chequamegon case, the Department of Justice, on behalf of the Forest Service, stated (Cheq. Brief 62):

> A natural forest by contrast is one which is both unexploited and unmanaged, where the succession of species is allowed to proceed to a "climax" state with only the interruption of natural catastrophe.... With the exception of the 10,818 acres [4378 hectares] designated by Congress as wilderness, and certain other areas preserved from management for other reasons, the Chequamegon National Forest is not, and will never be (absent action by Congress to create additional Wilderness areas to alter the multiple-use mandate) a natural forest.
>
> While we concur that the regulation in question [36 C.F.R. § 219.27(g)] requires the Forest to consider the natural forest in its planning, it would be *irrational to read into it a requirement that the Forest actually plan for the achievement of a goal contrary to its essential multiple-use mandate.* (Emphasis added)

This position is shocking for anyone seeking to preserve natural ecological processes in any portion of multiple-use lands. In the view of the plaintiffs in the suit, this could not be the law for a series of reasons:

- The diversity regulations are not limited to Wilderness Areas, for which there is a separate regulation (36 C.F.R. § 219.18). If the Committee of Scientists intended that diversity was only a concern in Wilderness Areas, the language of Sections 219.26 and 219.27(g) would have been lodged in the wilderness regulation.

- Even more clearly, the diversity language in the statute plainly contemplates that diversity is a part of the multiple-use analysis, not inconsistent with it, and is not restricted in its application to Wilderness Areas rather than to the whole forest, as the Agency suggested.

- Nothing in the Multiple-Use Sustained-Yield Act of 1960 (16 U.S.C. §§ 528 et seq.) reflects any such animosity to natural ecological function. Even if the only way to provide diversity would be to halt the reduction of mature forest to small patches by establishing DMAs, the definition of "multiple use" recognizes that "some land will be used for less than all of the resources" and that multiple-use management will be conducted "without impairment of the productivity of the land" [16 U.S.C. § 531(a)].

- Is the Agency saying that there are no vestiges of ecological function or remnant unique elements of biodiversity (e.g., populations of old-growth species) in the actively manipulated lands that could be preserved or enhanced? Is the ecosystem principle of 36 C.F.R. § 219.1(b)(3) now moot in Wisconsin because the presettlement forest is virtually all gone? Where is the authority (or public support) for this extreme position? Certainly this disdain for maintaining and restoring ecosystems is contrary to every public relations statement of the Agency and flies in the face of the broad statements of purpose set out by Congress and the Agency itself.

The conservation plaintiffs urged the court to reject the Forest Service's internal cultural view of multiple use as ubiquitous manipulation and likewise to reject the view that once land had been cut and was no longer in its presettlement condition, the ecosystem and diversity language of NFMA and the diversity regulations no longer applied.

Summary

Multiple use must accommodate lands on which preservation and restoration of natural patterns of disturbance are a first priority if it is to merit survival as a valid general theory of forest management, capable of addressing biodiversity concerns among other objectives. A definition of multiple use that abhors natural ecological structure and function is fundamentally at odds with the Agency's responsibility to safeguard the biological heritage of the lands entrusted to it.

The Wisconsin cases give us a detailed view of how diversity was addressed in the mid-1980s and which actions the Forest Service is and is not willing to undertake in order to protect biodiversity in the future. As inadequate as the Nicolet and Chequamegon initial biological analyses in the mid-1980s were, the Agency's refusal to reject that work and embrace the appropriate analytical tools that would allow for an informed debate about the biological fate of the forest is even more troubling. As this struggle continues, the Agency insists on maintaining high intensities of landscape disturbance, apparently in the belief that if species loss reaches a crisis level, it can simply "provide habitat capability needed to support populations of selected species at selected levels" (USDA Forest Service 1991). Bear in mind that the ANPR proposal, which showed so little evidence of ecological awareness, much less a willingness to make ecological principles central to performance of the diversity mandate, occurred in February 1991, some 43 years after Leopold's death and four-and-one-half years after an extensive written filing had challenged the Agency's competence in ecology.

Only when the Agency realizes that nature, not the Agency, builds forests, that the effects of managing certain species at "selected levels" cannot be isolated from the remainder of the biota, and that natural ecological processes cannot be addressed solely in local project terms, will there be the requisite viewpoint to begin addressing biodiversity in an ethical and effective manner. If this development has not occurred in the more than 40 years since Leopold, how can we expect to bring the Agency to that viewpoint in the foreseeable future?

14

Dominant-Use Zoning and Wildlands: Forestry for the Twenty-First Century

Pinchot's response to the massive clearings at the turn of the century was a simple directive. If we restrained ourselves from greedy, wholesale takings of the tree cover and invested in judicious replantings, we could consider ourselves responsible stewards and retain all the benefits of our forests. As the century closes, we have not outgrown the need to remind ourselves of Pinchot's basic conservation message. Even this modest level of restraint is not observed on many private forests and public old-growth lands when to do so would sacrifice short-term profits (Egan 1993).

But in the vast majority of public forests, Pinchot's resourcism has become a dogmatic view that tree cover constitutes the relevant feature of forest habitat, humans can actively manipulate tree cover to achieve desired populations of any other selected species, and intentional movement toward wild forest conditions is neither a wise nor legitimate multiple-use management option.

Leopold realized that wild lands contained complex biological systems not readily duplicated by human cultivation, and that an ethical approach to stewardship must account for the full ecological content of the land, not just those elements of it that might be useful to provide food, shelter, or sport for humans. Yet when directed by Congress to develop Forest Plans for issuance in the mid-1980s, and when specifically directed to address the problem of biodiversity, the Forest Service did so in Pinchot's terms, not Leopold's.

In this final chapter we set forth several proposals to uproot Pinchot's resourcism and its technocratic belief that humans can achieve precise desired results in forest content. To do so we must replace Pinchot's narrow view of "scientific forestry" with a forestry that is based on a new and broader set of scientific disciplines. And we must not fall into the trap of letting our relatively new knowledge acquired in the last two decades give us a sense of arrogance about our management proposals in the way that Pinchot's philosophy imbued generations of foresters with overconfidence and disdain for the views of those with different training. Rather, we must take from the new sci entific knowledge its real message, which is the understanding of the vastness and complexity of the ecological systems upon which the forests—and ulti- mately all species, including humans—depend for their sustenance.

Our diversity policy consists of several elements, each of which will be de- scribed in greater detail in the remainder of this chapter:

1. *A new scale for forest management.* We must preserve and regenerate large unfragmented blocks of late successional forests. In doing so, we explic- itly reject a small geographic scale (the "prevailingly local" approach) as the exclusive, or even primary, level on which to make management de- cisions for biotic protection and replace it with explicit recognition of the roles of landscape processes.

2. *The elevation of science.* We must elevate the use and weight accorded to scientific facts, testable hypotheses and predictions of long-term effects, as the primary basis for developing strategies to protect biodiversity. In doing so, we demote or eliminate unfounded assumptions and cultural factors, while promoting full disclosure of environmental consequences as predicted by existing credible scientific evidence.

3. *A new definition of wilderness.* We must redefine wilderness to be selected, designed, and managed from a biotic perspective. Recreation and aes- thetic criteria will remain legitimate considerations, but they will not suf- fice if we hope to achieve the wild conditions that will support many of the most threatened biological communities.

4. *Reinventing multiple use to incorporate dominant-use zoning.* Multiple use must be restated to openly embrace the allocation of highly incompatible uses largely to different areas of the forest, rather than assuming that they will doubtless complement one another on virtually every hectare over time. If the reality of dominant use, whether by timber harvest or old-growth forest protection, is not accommodated as legitimate within multiple use, then multiple use will become a historic relic unable to transcend its many decades as a code word for timber primacy.

5. *Expanding legal protection for biodiversity.* We must argue for an expansive interpretation of current laws to achieve the protection of ecosystems in- tended by Congress and enact greater protections for ecosystems which advance our commitment to the environment beyond the single species approaches employed to date.

6. *Managing with an understanding of the limits of our knowledge.* We must expose the hubris of management for "selected species at selected levels"

as directly at odds with the Land Ethic of Leopold and insist that persistent landscape manipulation not continue on lands of high biotic value in the face of considerable ignorance of the long-term biological consequences of such management.

The Need for Regeneration of Large Unfragmented Blocks of Late Successional Forests

The first prescription for a new diversity policy is to demand that our federal agencies achieve preservation and regeneration of large unfragmented blocks of late successional forests. The specific environmental conditions in each region in the country will dictate the specific size and location of such areas, as well as the degree of human-introduced disturbance appropriately permitted in them (Chapter 11). We use the term "preservation" to mean the existence, to the extent practicable, of forest conditions to which native species are adapted and in which they can continue to evolve without substantially accelerated rates of extinction.

In the West and Pacific Northwest, there still remain opportunities for preservation consistent with these goals. In the public forests of the eastern and midwestern U.S., we may be able to combine remaining old-growth and mature forest tracts, Wilderness Areas, and Research Natural Areas with additional lands designated for restoration into large blocks needed to improve resistance to the adverse biological consequences of fragmentation described in earlier chapters. Remaining islands of old growth and persistent populations of interior species will act as important inocula for this restoration. By carefully zoning sufficiently large blocks using these elements, we can allow restoration to progress, primarily through natural succession, toward a state of greater ecological integration and integrity. Given the patchwork quilt of ownerships in the eastern half of the country, these Diversity Maintenance Areas will need to be made up from all types of ownerships: federal, state, county, and private land.

The criteria for DMAs were discussed in Chapter 11. Present ecological conditions of the upper Great Lakes forests suggest that areas large enough to maintain functional integrity over the long term (at least 200 years) should be in the range of 20,000 hectares (50,000 acres), at a minimum, to a more desirable 60,000 hectares or more. The appropriate size of DMAs in other areas of the country must be determined by thorough study of the relevant ecology, but will likely be of this size or larger. Some have suggested that as much as 50% of the land base will need to be returned to large contiguous chunks of naturally disturbed lands (Noss 1992; cf. Leopold 1936). The Northern Rockies Ecosystem Protection Act proposes protective designations for over 8 million hectares in the public lands of the Northern Rockies in five huge blocks and connecting corridors (Bader 1992).

The future condition of DMAs will no doubt differ in some respects from the presettlement conditions before forest destruction and the effects of persistent fragmentation spread across the landscape. Yet such large blocks of the future will exhibit unique elements of biodiversity and a much more robust "immune system" protecting that diversity from the hostile forces of disturbances common outside of these blocks. And in these blocks there will be conditions much more closely resembling the evolutionary context that existed prior to the ongoing spasm of extinction brought on by our drastic alteration of the landscape in the last few centuries. If we do not allow these blocks to be preserved, restored, and sustained, the detrimental area, edge, and isolation effects discussed in Chapters 5 and 6 will continue unrelentingly to reorder the biotic content of our world—favoring some species and gradually, almost imperceptibly, extirpating others.

A close corollary of the need for biotically sufficient blocks of forest is the recognition that "prevailingly local" approaches to management should no longer be accepted as a satisfactory answer to the ecological processes that underlie species loss. The "prevailingly local" approach can most readily be seen in well-intentioned attempts to protect threatened, endangered, and candidate listed species on the basis of site-specific mitigation steps. While these measures are often indispensable in avoiding outright takings through destruction from machinery or dramatic local changes in habitat conditions, because of limited resources and limited knowledge, they cannot be relied on to effectively protect communities over the long term. If local measures were adequate, we would not see the continuing population declines and progression of more species to listed or candidate status. The "prevailingly local" approach must be overcome not only in preparing individual Forest Plans, but in determining Regional and Agency-wide goals in the RPA (Forest and Rangeland Renewable Resources Planning Act of 1974, 16 U.S.C. §§ 1600 et seq.) process.

These challenges are well within the capability and interest of many Forest Service personnel, in whom often lies suppressed a great capacity to enact Leopold's Land Ethic. We see this in their decision to work in the forests, in the formation of the Association of Forest Service Employees for Environmental Ethics (AFSEEE), and in many off-the-record conversations with Agency staff who acknowledge the biological costs of continued fragmentation. The Forest Service already has ample statutory authority to manage for biodiversity. Instead of being preoccupied with forestalling the mythical timber famine, the Agency should exert leadership in addressing the clear and present danger to biodiversity.

The primary measuring stick of success in protecting biodiversity should be the extent to which ecological composition, structure, and function are restored and maintained. Other traditional units of measure such as species lists, extent of single-species mitigation plans, new land acquisitions, funding for recovery programs, the extent of clearcutting, the number of Research Natural Areas, or the acreage in so-called old growth (if defined as tree stands past normal harvest age) are of considerably less value in determining the long-term fate of biodiversity.

E. O. Wilson, in his book *The Diversity of Life*, makes a principal element of his formula for saving biodiversity a call to "restore the wildlands" (page 340):

> Yet there is still time to save many of the "living dead"—those so close to the brink that they will disappear soon even if merely left alone. The rescue can be accomplished if natural habitats are not only preserved but enlarged, sliding the numbers of survivable species back up the logarithmic curve that connects quantity of biodiversity to amount of area. Here is the means to end the great extinction spasm. The next century will, I believe, be the era of restoration in ecology.

This is Leopold's message with greater evidentiary support and greater urgency.

Without discarding local protections for truly rare species, scientists, activists, and land managers must realize that the question of ecological integrity has now been called in unmistakable terms. No plan or EIS for a large area of public land can be considered complete, even competent, in its analysis without facing the question of the need for Diversity Maintenance Areas and the environmental consequences of management that would fail to provide for the preservation or regeneration of such large blocks of forest.

The Elevation of Science as a Guiding Force in Forest Management

The elevation of science in forest policy should consist of four components: (a) broadening the set of disciplines applied to the problem of forest management; (b) giving scientists the respect and clout in Agency decisions that their information warrants; (c) demanding of scientists that they forcefully articulate the differences between their views of biologically sound management and proposed Agency decisions; and (d) disclosing the predictions and advice of those scientists to the public in the planning and environmental analysis (EIS) process.

Although Pinchot often termed his revolutionary approach to forestry "scientific forestry," there is now a basis for applying a much broader set of scientific disciplines to the understanding of forests. Systematics, population biology, population genetics, and community and ecosystem ecology are all indispensable in comprehending forests and should be applied before specific management decisions are taken. Regrettably, these disciplines were not applied to the planning efforts in the Wisconsin National Forests, and the Agency refused to directly address the scientific issues raised in the appeals.

Our view that science must be elevated in the management of public forests does not suppose that conservation scientists will agree with us or with one another. What is more important is that the public benefits from a frank scientific debate. Without the scrutiny and challenge of open debate, the best scientific evidence may be ignored and opposing technical viewpoints unheard.

After relevant scientific disciplines are allowed into the policy arena, we must then ensure that the opinions of the scientists will be accorded the respect necessary to make a difference in Agency decisions. Much has been written about Forest Service personnel who tried to apply their scientific knowledge only to be censored, reassigned, or forced into retirement.

The Forest Service has a Research branch, which is separate from the National Forest System, the branch charged with policy and actual management of the National Forests. Unfortunately, the input from Forest Service Research is heavily discounted by line officers when arriving at decisions for management of the land. Letters from Research are found in the appendices of planning documents, but their opinions rarely directly influence specific management decisions. Line officers from the National Forest System often discount the importance of Agency Research with comments such as "but he was only from Research" (Hathaway 1992, pers. comm.) or, alluding to comments from the head of the Research branch on the authors' DMA proposals for the Chequamegon National Forest, "sketchy advice based on sketchy info" (WFCTF 1986).

The Agency has long sponsored scientific work of importance, yet a clear distinction remains within the Agency between scientific thought and management decisions. This dichotomy hinders implementation of protections for biodiversity. It is difficult to feel encouraged about efforts at reform within the Agency when a senior career officer, Region One Regional Forester John Mumma, can be removed from office in part for sustaining the position of his forest supervisors that many timber harvest levels in that Region were excessive and would violate the constraints of the National Environmental Policy Act (NEPA). Biologists also were dismayed when Forest Service botanist Karen Heiman was abruptly fired in March 1991, apparently in response to being an active advocate for protecting rare plant habitats (St. Clair 1991). Less conspicuous, of course, are the daily indirect threats and other pressures that many Agency personnel experience to conform to timber-cutting policies. Small wonder that many biologists remain silent for fear of losing their jobs or because they see that the only road to promotion requires that they remain silent and try to live with Agency policies or actions.

Efforts to stifle scientific input or punish professionals for applying their expertise and judgment rightly engender the indignation of the wider scientific community, as evidenced in a letter from academics to Vice President Gore (Bolle et al. 1993). Indeed these efforts represent both a suppression of scientific understanding and a blow against scientifically based decision making—the heart of NEPA and NFMA. To purposefully ignore or suppress scientific understanding as we develop plans to manage the public forests therefore represents both a legal and a scientific subversion of policy decisions. This contributes greatly to the public's distrust of the Agency and a growing cynicism that many National Forest Plans and FEISs resulting from such subverted processes are fundamentally political documents reflecting a heavy timber bias.

Some feel strongly that the federal agencies, including the Forest Service, are already in the midst of a remarkable period of self-evaluation and redirection which will elevate scientific discourse in the decisional process. For example, recent reviews of policies within the National Park Service (Risser and Lubchenco 1992; National Academy of Sciences 1992) are unanimous in calling for greater emphasis on ecological inventory and monitoring, understanding ecosystem processes, and cooperating with surrounding landowners to facilitate responses to trans-boundary interactions. In Wisconsin, one tangible and important result of the National Forest cases was the Chief's order in his appeal decisions that the two Forests convene a "committee of experts" to consider the appellants' concerns. This order resulted in a Scientific Roundtable on Biological Diversity, which issued an impressive report describing in detail a variety of management changes that should be implemented and researched in order to protect biodiversity (Crow et al. 1993; Waller 1993b).

The federal Agency most directly charged with environmental protection, the Environmental Protection Agency (EPA), has also recently begun to show an increased level of interest in biodiversity issues. In 1990, the EPA Science Advisory Board reported that the EPA was spending disproportionately too much attention on localized and highly visible issues such as toxic wastes and too little attention to the more subtle, but ultimately more significant, biological issues surrounding habitat loss and species endangerment. Thus, there is some evidence that the federal government is slowly moving toward a more integrated and inclusive set of policies, grounded on scientific principles, to protect biodiversity.

Our own experience, however, lends a note of caution to those seeking comfort at the advent of Forest Service reform on biodiversity policy. For years the Agency has had ecosystem and diversity language expressly in its governing statutes and regulations. We have sat through well over 100 meetings where "commitments" to biodiversity protection were voiced. Yet the history of the Wisconsin cases remains strong evidence of the Agency's beliefs that it can meet such a commitment to biodiversity through a prevailingly local, single-species approach. It is up to the scientific community to make the case that such piecemeal approaches will no longer suffice.

As E. O. Wilson has said, we must begin treating the problem of biodiversity protection as if it were national defense or public education (Wilson 1992). How can we ask the public to give biological science that level of interest and support (such as through a National Institutes for the Environment) if the scientific community itself does not think its results are sufficiently important to be an indispensable component of the management of our public lands? In the words of Thomas Lovejoy, "It is up to science to spread the understanding that the choice is not between wild places or people" (Lovejoy 1980).

If the Agency defined ecosystems in full biological terms, not simply in terms of tree cover, soil conditions, or current species lists, this would immediately broaden the debate about the effects and trade-offs of various management choices. And if the Agency frankly set forth current scientific opinion on the importance of large blocks of forest subject to natural disturbance regimes, the importance of management choices would be plainly before the people.

To further hasten that day, let scientists join with local citizen activist groups interested in each and every National Forest to present a set of basic conservation biology materials to each forest, before and during major efforts at forest planning. How then could the Agency avoid addressing the environmental costs of forest fragmentation?

Redefining Wilderness

Whether called wildlands, wilderness, DMAs, biological reserves, or 9.3 Management Areas, we must find a way to select and designate lands in which our principal goal is to restore and preserve ecological integrity. Leopold understood that in wildness was ecological integrity, and it is his conception of wilderness that must be at least coequal with that of the backpacker or canoeist seeking pleasing vistas, solitude, and a sense of wildness bred of risk to human life. Human values pass legislation, but if human values now include real protection of biological diversity, might not the biological value of wild areas become a preeminent notion of wilderness?

The merits of associating ecosystem protection with wilderness lie in their very close interconnection in fact, the centrality of a hands-off policy for achieving biotic health, and our proven willingness as a society to take strong legal measures to protect lands of exceptional value, called wilderness, for nonmarket reasons. These hallmarks of wilderness would well serve biodiversity. It may be that to obtain sufficient wildlands for protection of biodiversity, we need to consider some human intrusions (such as snowmobiling or subsistence logging for firewood, as in our DMA proposals) that pose little threat to biotic values even if they vary from traditional, aesthetically based notions of wilderness. Similarly there may be some traditionally permitted activities in Wilderness Areas that create disturbances which are incompatible with biodiversity protection. This may mean different classes of wilderness—some for aesthetic purposes and others for biotic purposes—with a great deal of overlap and mutual support.

To carry out the creation of greater wildlands with a biotic purpose, we should reassess all lands under public ownership with a new ecological and long-term perspective, regardless of whether some lands might have been too heavily roaded to qualify for RARE II wilderness consideration. Roads can be "rolled up" with giant-toothed rippers or, like existing wildlife openings, simply abandoned to natural regenerative processes. All public lands should be reevaluated not only on the basis of current condition but in a landscape context, with a view toward what those lands might contribute, based on their size, location and relationship to neighboring lands of important biological content, to ecological integrity centuries in the future.

In many locations, the premises of RARE II evaluation may be a good deal less relevant to the long-term ecological integrity of public forests than criteria grounded in biogeographic principles. While lack of roads usually has meant lack of human-caused landscape disturbances of other types, it is primarily in the light of long-term biological diversity that we must redefine the concept of wilderness and our reservations of specific wildlands. We may take inspiration from Leopold's biotic notion of wilderness in making this case.

Reinventing Multiple Use to Incorporate Dominant-Use Zoning

It is doubtless impossible to debate the meaning, design, and location of wildlands without facing what has been that great counterweight to wilderness: traditional multiple use. We should avoid the temptation to cast wildlands and multiple use as inevitable opposites, locked in a struggle that paralyzes our forest policy as wild species and communities continue to suffer. Multiple use need not become a relic of twentieth-century forest policy, incapable of coping with the biological demands of late twentieth and early twenty-first century conservation biology.

Rather than rejecting multiple use outright, we emphasize the need to reject the traditional view of multiple use, as it has been interpreted by the Forest Service and other forestry bureaus at the state and county level, to mean that logging categorically enhances wildlife and other values (Chapters 8 and 9). Multiple use need not necessarily be redefined in the law, which allows wide latitude to consider nonconsumptive uses and the protection of the diversity of plant and animal communities. Instead, we should accept the fact that certain uses have in fact dominated many multiple-use management plans and abandon the myth that all uses can coexist at some point on the same acreage. These points have long been evident to the ordinary visitor to the public lands and were succinctly recorded by the economist Clawson (1975) almost 20 years ago.

Advances in ecology over the last 20 years corroborate the case that timber is a more dominant use than even its critics have usually asserted. Biological impacts extend to areas adjacent to those actually logged, and areas continually fragmented have little opportunity to recover their ecological structure and function. Society must bluntly face the fact that logging is currently a dominant use in so-called multiple-use lands through its pervasive effects on the surrounding landscape. Only a similarly dominant use for conservation over large areas can successfully protect biological diversity over the long term. By demonstrating these two sides of the same coin, conservation biology leads us to accept dominant-use zoning plans as the relevant and appropriate subset of management approaches which should now be considered.

The Agency's interpretation of multiple use, grounded in the philosophy of Pinchot, finds the Forest Service reluctant to consider passive restoration or large reserved areas off limits from commodity production. These low-cost, low-impact proposals have been falsely characterized as "single use" or "withdrawals from multiple use," despite their value for watershed, recreation, and other uses. Critics of these ideas believe that an area cannot be considered multiple use if it does not allow timber harvesting. But if all potential uses must be available at some time on each tract, when and where in the forest do those plants and animals that need large unfragmented blocks of mature forest get their turn?

The Agency's cultural rejection of wild conditions as a management goal on managed forestland has not been adopted by Congress. On the contrary, "the establishment and maintenance of areas of wilderness are consistent with the purposes and provisions of" the Multiple-Use Sustained-Yield Act of 1960 (16 U.S.C. § 529). Federal agencies must be pressed on this point and challenged to define multiple use, in practice, as including lands for which active and passive management for restoration of natural forest conditions is a primary objective. Wildlands should not be perceived as apart from multiple-use lands, set aside to lie idle in some useless and wasteful state, but rather to be a critically important part of an appropriate and integrated management plan for large forested areas. If multiple use cannot be redefined to allow for the inclusion of wild conditions, then the job of managing commodity-dominated lands will need to be separated from the task of protecting native biodiversity, by cleaving the Agency in two or more parts or reassigning responsibility for biodiversity to another agency.

Expanding Legal Protection for Biodiversity

To protect biodiversity, we must make the most of the laws we currently have which address this problem, albeit imperfectly. This will entail protecting those provisions already enacted or promulgated from weakening amendments or repeal and promoting the reading of those laws in light of the new knowledge we are gaining from conservation biology. Second, well-considered amendments to current laws should be proposed that would more explicitly empower agencies to accomplish the tasks of protecting large blocks of intact ecosystems as well as the other tools of biodiversity conservation.

Although most discussions of legal protections for diversity begin with ESA, our focus on the National Forests prompts a first look at NFMA and its regulations. The appropriate legal foundations for a new diversity policy should be grounded in the Agency's definition of "diversity" (36 C.F.R. § 219.3), its statutory duty to "provide for the diversity of plant and animal communities" [16 U.S.C. § 1604(g)(3)(B)], and the planning goal of a "natural forest" condition [36 C.F.R. § 219.27(g)] found in Forest Service regulations and discussed in Chapter 13.

Although the NFMA diversity language contains qualifiers regarding multiple-use objectives, all forest policy must currently be carried out within the multiple-use framework, and this is not reason enough to completely discount the diversity mandate in NFMA. Indeed, under that statutory authority, the Agency's regulations have generated both a diversity definition which fits well with current concepts of biodiversity and a conceit that the natural forest is a valued state of forest condition—an extraordinary repudiation of Pinchot's disdain for primary growth. The natural forest language is the only place in current law to our knowledge where one can say the Land

Ethic of Leopold—the vector of management back to a wilder condition—is found in explicit terms in the law regarding forest management.

In Chapter 13 we outlined attempts the Forest Service has made to interpret multiple use as at odds with the natural forest condition and, further, to strike the natural forest language from the regulations. This language must be safeguarded as the first light of a new day in land management—a reversal of Pinchot's dislike for the wild forest and a turning toward an appreciation of its biological significance.

Another source of legal foundation for the new diversity policy is found in that much maligned statement of our national environmental policy, the National Environmental Policy Act (NEPA). The choice of a new diversity policy by the public and Congress will in large part be a function of future Congressional hearings, public debates, citizen comments, and Environmental Impact Statements. As we saw in Chapter 13, NEPA's current language not only elevates scientific evidence in the decision-making process but also emphasizes the environmental effects on ecosystems in a key regulation underlying the statute, 40 C.F.R. § 1508.8. One element of a new diversity policy that may draw on current law is using the NEPA public participation procedures, whether one is a member of Agency staff or a member of the public interested in protecting biodiversity, to fully disclose the implications of modern conservation biology.

We must also resist attempts to weaken the Endangered Species Act. As of this writing, there is much discussion about amendments to the ESA during the reauthorization process. Perhaps the latest round of that process will be concluded before this book is published, but there will be future reauthorizations and thus chances for improvement or regression. The ESA has proven to be a formidable obstacle to extirpation in cases of popular species that can obtain listed status, and the Act is much more clearly mandatory in its limitations on development pressure, public or private, than the more discretionary language guiding the Forest Service.

Yet the ESA has not prevented additional listings, and biodiversity remains increasingly threatened. The ESA's mechanisms were invented under the same false premises that were articulated by the Forest Service in its 1986 Forest Plans for the Wisconsin National Forests, namely that diversity is flourishing, ecosystems are largely functioning quite well, and we only need to attend to a few species with particular, localized habitat problems. It was, and is, unrealistic to expect a law designed in that context to adequately handle the flood of species in need of help as their basic support systems become increasingly dysfunctional (Winckler 1992; Kohm 1991).

What appear as funding shortfalls, lack of appropriate priorities, inadequate habitat acquisitions, and woeful delays in listing decisions could be readily cured if it were not for the rising tidal wave of species demanding protection. The ESA is akin to a small soup kitchen trying to help a long waiting line of disaster victims, while the basic infrastructure of power, water, and other essential services is no longer functional. We do not proclaim the soup kitchen a failure and allow those in most immediate need to starve; rather, we try to keep it going even as we rebuild the community so that it may again have the necessary structure and function to support life. Particularly given the long time span over which ecological reconstruction will work its beneficial impacts, we must retain and adequately fund the current provisions of ESA protecting diversity under a species-by-species paradigm, even though we recognize that such methods are not a panacea for the continuing losses in biodiversity and that new statutory language must be adopted which goes beyond that species-by-species approach. New legislation that would address these problems could take a number of forms.

The ESA (or similar future laws) could require that any designations or considerations of habitat address viability of the habitat itself with regard to all biotic inhabitants

Currently, the ESA allows determinations made with respect to habitat to consider only the needs of the listed species. If the species has small area needs or is very localized, listing does little for the surrounding ecosystem or its inhabitants. Even a highly mobile species with a large home range is not an ideal candidate because management can become skewed to maximize conditions for the listed species at the expense of the ecosystem as a whole, as it has for the timber wolf in northern Wisconsin (Chapter 12). In this respect, the ESA may work against the overall land health of the area by prescribing emergency steps that were not present in that ecosystem when the species now of prime concern enjoyed much higher population levels.

The Act would be improved by requiring that habitat identified under the Act pass its own test of viability in broader ecological terms, as a functional unit with regard to its biota as a whole, not just be deemed sufficient in terms of the food, cover, etc. needed by the listed species. This could be defined by reference to the frequency and distribution of indigenous ecological processes or to known or predicted ecological requirements of all biotic inhabitants, as represented by numerous, well-chosen indicator species representative of a broad spectrum of taxonomic groups and ecological guilds.

The teeth in this approach would be that once the habitat was identified in this way, adverse modification of that habitat would itself constitute an action adverse to the interests of the Act (ecological integrity *per se*) and thus be barred. This treats the presence of a listed species as one of perhaps many important indicators of problems for a given ecosystem. Such an approach should also lead to the designation of larger critical habitat areas as principles of reserve design dictate the needs of the ecosystem in which listed species reside. In the end, we would be protecting habitats themselves that passed tests of conservation biology, rather than specifically promoting single or a few species at a time.

To be a viable approach politically and economically, this idea may have to be limited to habitat designations or considerations on public land. It may be unrealistic to require that the developer of a small plot who is seeking acceptance of a Habitat Conservation Plan needs to compensate for the loss of this "postage stamp" of butterfly (or bird, or turtle) habitat by providing a functional ecosystem. On the other hand, when habitat is designated within a National Forest for something like the northern spotted owl or the timber wolf (critical habitat in Minnesota, but a national and state recovery plan in place for Wisconsin as well), the habitat that is described as critical for such species should pass a conservation biology test of long-term viability for the entire community in which the listed species resides, not just the listed species.

All federal agencies could be required to provide for diversity of plant and animal communities

This suggestion would add a specific diversity requirement to ESA or other federal laws to match (and hopefully improve upon) the diversity requirement now applicable to the Forest Service. Perhaps the best argument in favor of this approach is that it is already something that Congress has adopted for one major agency, and it should be expanded to cover the actions of all agencies through the consultations that they must undertake under Section 7 of the ESA. The value of such a provision would of course depend on the meaning the federal courts ascribed to the statutory language presently in NFMA.

We would suggest taking out the multiple-use constraints now in the NFMA language, as these would not apply to agencies generally. And if such language were added to other statutes, a statutory definition of "diversity of plant and animal communities" would probably also be in order. We would suggest that the definition in the current Forest Service regulations (36 C.F.R. § 219.3) matches well with gamma diversity. One problem presented is that the current Forest Service regulation is tied to the "area covered by the land and resource management plan." We would need to specify some area for the definition to make sense ecologically, but the relevant area may vary depending on the context in which the definition is used.

Create a federal "program" for biodiversity

For several years, Congress has entertained legislation aimed at creating a biodiversity program to oversee the analysis of the biodiversity problem in this country and to which all other Agency actions would somehow be tied. This could be developed under the "program" to be prepared by the Secretaries of Agriculture and Interior, under Section 5 of the ESA, but which as of this writing has not been established.

Such a program would of necessity reach not only listed species but those not yet endangered or threatened, and set out a plenary federal strategy to protect biodiversity. While such a program would not necessarily lead to quick implementation of protective measures on federal lands, it would provide, through the efforts of local activists, a stronger long-term footing for biodiversity protection than the piecemeal development of diversity protection measures in individual National Forests or other public lands.

Such a comprehensive program could well need new specific agency support, both for inventorying the ecological attributes of the country (through the proposed National Biological Survey) and through substantive proposals for ecosystem protection that might be generated through a new National Institutes of the Environment (NIE), similar to the National Institutes of Health. While the mission of this latter entity would clearly be research and not policy or management, it is likely that the integrated efforts of one or more institutes within the NIE would focus scientific opinion more efficiently and effectively on policies of the federal government. Agencies charged with managing federal lands would, of necessity, seek to incorporate the findings of such an institute in order to maintain the confidence and budgetary support of Congress.

Land acquisitions should be expanded, with priorities determined by biogeographic considerations

Additional public land is needed to meet the biogeographic requirements of diversity protection, especially in the eastern biomes and along watercourses in the West. This should be complemented by acquiring partial interests in land, such as restrictive covenants or easements, to afford protection against disturbance without necessarily holding complete ownership (as is often used for aesthetic purposes in scenic riverways and similar situations). Such acquisitions must be targeted, both at the gaps in current species composition in native diversity and on the basis of biogeographic criteria which address the long-term viability of ecosystems. Tax incentives at the state and federal level could create economic incentives for the management of private lands to complement these objectives.

Laws should be adopted which expressly protect ecosystems

More sweeping proposals have been made that would identify endangered ecosystems, rather than just species, and prohibit the taking of rare habitat, regardless of any immediate proof of individual species impact (Noss 1991d). While this approach properly focuses on habitat, it could be misinterpreted under intense political pressure to cause the federal government to identify only a few specimens of key habitat in a region, without necessarily creating an overall context of land management within which such specimens can remain viable and dynamic over time. This could result in a collection of a few ecosystem specimens, conceived of and managed as static types, perhaps no more viable than the small current specimens kept under the name of Research Natural Areas. The key to this type of proposal, as with critical habitat under ESA, or diversity management under current language in the NFMA, is that the habitat must be tested under the principles of conservation biology for its viability, rather than simply its current content.

Another approach to a broader habitat designation scheme would be to classify all U.S. lands into areas that were afforded different degrees of species and habitat protection (cf. the zoning schemes of Harris 1984 and Noss 1987b). Perhaps Class I areas would be national parks and wilderness, coupled with surrounding areas identified as DMAs under biogeographic principles. In these areas, very high standards of species and ecosystem protection would be afforded. In Class II areas, often concentric to Class I areas, lower standards would be met and certain types of development or use permitted. Class III areas would afford an even greater degree of certainty to development interests. Such zoning represents a logical combination of Harris's and Noss's ideas with our Dominant-Use Zoning and DMA proposals here.

Many states currently use such a classification scheme for surface water protection. The different classes are used to regulate current discharges as well as new applications for "degradation." The Clean Water Act has an "anti-backsliding" provision, which bars reductions in discharge controls as well as any lowering of water quality. For example, in Wisconsin's water quality program, "exceptional" surface waters are pristine in quality and not allowed to accept any new or increased sources of water pollution. Other less stringent classes require various degrees of "anti-degradation" analysis before new pollution can enter the water (Chapter NR 207, Wisconsin Administrative Code).

Our designated federal Wilderness Areas do not enjoy parallel safeguards against biotic degradation. Yet in the small Wilderness Areas designated in the eastern United States there is continuing degradation of the biotic content through edge, area, and isolation effects. Indeed, a strong case can be made that the protection of surface water resources in pristine areas of the

Great Lakes states is far greater than that afforded to the biotic resources of neighboring terrestrial habitats.

In addition to the anti-degradation concept in surface water programs, some federal environmental programs are quite blunt in adopting a strong, and very expensive, principle of insisting on restoration of the environment. Cleanups of groundwater at Superfund sites are guided by the notion that no polluted aquifer is to be "written off." Even when drinking water supplies are not directly affected, Congress and the enforcing agency (U.S. EPA) have readily accepted a very stringent and costly standard. Very local aquifer pollution often requires tens of millions of dollars for "pump and treat" groundwater remedies. One must ponder what might be done with such funds in the context of buying timber production rights in perpetuity, on a present value basis, and turning over the money to local communities for tax support, or job relocation or retraining, while protecting regionally significant tracts of regenerating old-growth forest.

If no aquifers are to be written off as spoiled by human activities, why should our public lands and their ecosystems be allowed to be degraded without a similar outcry and protection? Once the legitimacy of the conservation biology principles espoused in this book become accepted criteria for forest management, we may be able to progress to bolder strategies that demand that biodiversity be given the kind of stringent protections afforded environmental qualities of greater familiarity and more traditional attention in our society, such as water quality and solid-waste disposal.

The Importance of Humility: Managing in the Face of Uncertainty

Just as important as the need to press into action the knowledge we have gained from conservation biology is the need to appreciate from this learning just how complex are ecological processes and thus the vastness of what we do *not* know about natural systems.

In late 1992, six years after the Forest Plan appeals, the authors critiqued an individual timber sale proposal in the Nicolet National Forest. In the critique, we stated our concern that the Agency "did not know the environmental consequences of its proposed cutting." As usual, the Agency staff, here one of the most progressive district rangers to work in Wisconsin, protested this characterization in strong terms. Yet Agency representatives could not begin to answer questions we had asked six years earlier about the effects of this and similar projects on regional patterns of deer abundance in time or space, the population dynamics and habitat availability of certain orchids, or the presence and frequency of old-growth lichens or invertebrates. And what of ants, mycorrhizal fungi, and the countless other players in these ecological systems? Good intentions (so prevalent in the Forest Service) are not nearly enough: responsible management requires that we understand, and are accountable for, the environmental consequences of our actions.

If we scold the Agency for not understanding these complexities, we must readily admit that vast amounts of this information are unknown to us all. It is therefore critical to decide how management should be carried out in the face of this ignorance. We simply cannot go on blindly, drastically altering the physical conditions of the land, through weekly decisions in land management offices across our country. Our old great simplifications, assumptions, and hopes are no longer acceptable.

By increasing our awareness of the extent of our ignorance, conservation biology turns on its head the traditional characterization of management proposals for reserved areas as "radical" and for ubiquitous fragmentation through continued logging and roadbuilding as "wise, multiple use." The truly conservative management alternative is passive management for natural disturbance regimes, unless specific evidence (e.g., a local explosion of a highly competitive exotic species) suggests otherwise (Alverson and Waller 1993).

We continue to discover fresh ways in which anthropogenic disturbance and habitat fragmentation threaten vertebrate species. And the lion's share of biotic diversity consists of organisms that are poorly known: little-known species of fungi, lower plants, and soil invertebrates. This will likely remain true well into the next century. Many of these species will also be sensitive in their own peculiar and unsuspected way to the many disturbances associated with logging (Shaw et al. 1990). To fully understand the biology of these organisms and their interactions will, of course, require an enormous research effort. At least in the immediate future, most such research will remain undone.

Similar uncertainties have been faced in making decisions regarding the acid rain and global warming issues, where similar social and political questions arise. Here, in the presence of scientifically probable, but unproved, deleterious impacts, debates have emerged over the risks involved in postponing action until specific proof is available and over which interim decisions are most appropriate. Many scientists presume that science should play a dominant role in such policy debates, yet in these arenas, as with forest policy, public policies have usually followed their own dynamics (Ludwig et al. 1993). Even large and well-funded science programs, such as the $550 million National Acidic Precipitation Assessment Program, have often failed to effectively guide policy when political factors intercede (Loucks 1992).

While superficially similar to these other global change concerns, the issues surrounding biological diversity are unique. The causes and consequences of local extirpations, global extinctions, and other disruptions of ecological processes may be even more complex and unpredictable than changes in already highly complex physical systems such as the atmosphere. Need we concern ourselves with such uncertainty regarding the fate of biological systems? Haven't humans persisted and thrived for millions of years without this ability? In fact, such unpredictability has already resulted in many unpleasant surprises in the past (Elton 1958; Bratton 1975; Mooney and Drake 1986; Preston 1992; Ludwig et al. 1993).

These surprises will continue in the future. Collectively, they cast further doubt on our confidence that we can manage complex forests containing many sensitive components without significant or long-lasting impacts on their biota. Yet many approaches to forestry, such as Franklin's New Forestry Model (Chapter 11), justify continued, substantial biomass removals from the forests upon our ability to predict the ecological consequences of our actions. We are asked to trust foresters, that their technical training and confidence will suffice to ensure that all economic, aesthetic, and biological resources of our forests will be maintained despite appearances of soil erosion, failed regeneration of some species, and forest decline over many areas. Persistent controversies that have developed among many forest scientists and managers in recent years suggest that this confidence is misplaced. Trusting a new technocracy of foresters is a hauntingly familiar reminder of the generations of all-knowing foresters from the Pinchot legacy who brought us much of our current difficulties (Clary 1986).

A second way in which threats to biological diversity are unique is that some are irreversible. While we can often undo mistakes we make in designing industrial manufacturing processes that are wasteful, destructive, or threaten our own health, we have no way of remedying the loss of species. While the self-styled Wise Use movement may denigrate the importance of these losses and seek to attribute them to a lack of "adaptive ability" on the part of sensitive species, leading scientists and every major national and international biological organization have sought in recent years to warn the public of the magnitude of these impending losses and the many significant dangers such losses represent to humanity (Terborgh 1989; Ehrlich and Ehrlich 1991; Ehrlich and Wilson 1991; Wilson 1989, 1992). Ultimately, of course, the protection of species is an ethical and moral issue (Ehrenfeld 1978; Leopold 1949).

What kinds of decisions should we make when full information will never be available? As alluded to in Chapters 3 and 9, many land managers have approached their tasks with great confidence in the comprehensiveness of their training and with hubris regarding their ability to manage the land simultaneously for sustained production and the protection of biotic resources. While some scientists also express this trait (Salwasser 1987), and this view has been expressly pursued by the Agency in its proposed regulatory changes in early 1991 in its suggestion that it can manage for "selected species at selected levels" (USDA Forest Service 1991), many conservation biologists are considerably more humble in their confidence regarding the ecological sustainability of current practices (Ehrenfeld 1978, 1993). While it is the duty of each forest scientist and public forest manager to decide for herself or himself where to stand on the hubris–humility continuum, the public has the right to expect that all those charged with making and implementing forest policy will take their responsibilities seriously enough to handle our forests cautiously and with knowledge of, and respect for, their biodiversity.

Summary

To the Forest Service, E. O. Wilson's call to "restore the wildlands" is fundamentally at odds with sound or lawful management. Wild conditions, the "natural forest," and Diversity Maintenance Areas are not cognizable under the Agency's conception of multiple use. Yet many within the Agency and in the public are increasingly aware of the importance of biodiversity, the limitations (and occasionally confounding results) of species-by-species and site-specific approaches to species protection, the inadequacies of many existing biological reserved areas, and the growing importance of conservation biology.

While there is much more to learn about ecosystems, we must begin adopting protective measures based on our best current knowledge. That knowledge tells us that large blocks of relatively undisturbed forests are critical for the survival of many species and the ecosystem processes upon which they depend. To achieve large old-growth forest reserves, we should allow major portions of public forestland to be released from constraints presently imposed by active, manipulative management plans for commodity production and, simultaneously, encourage silviculture in other lands and promote what in the electric utility field has already become accepted as "demand side reduction" (i.e., seeing recycling and reuse as a biological imperative, not just a solid-waste disposal issue). We cannot expect forests to achieve long-term ecological land health if we do not permit the reestablishment of natural patterns of landscape disturbance.

The choice and design of locations for such forest restoration will only be achieved if we give the scientific community a meaningful role in the land management debate. Such elevation of science in the decisional process will help educate the managers and the general public about the importance of protecting large blocks of natural habitat and old growth.

We must also free ourselves of the rhetorical baggage of the past decades and rethink the purposes of our wildlands and the relationship of such lands to the other multiple purposes of federal land ownership. The vision of twenty-first-century forestry is still shrouded, and many competing models will no doubt be advanced. But central to any such vision must be a major component of lands whose management and form are driven by the overwhelming need to maintain the biotic richness and complexity of native interior forests.

Some of these ideas may require changes in the interpretation of our laws. Others may require new laws. Perhaps more important than any new set of scientific recommendations or statutory amendments, however, will be the development of a certain humility. We will come to respect wild conditions because they more closely approximate the evolutionary context in which the biota have arisen than any conditions that can be created through our active management.

With open scientific debate, aggressive application of the law, and leaders committed to conservation who are equipped with new conceptions of wilderness and multiple use, we may still see the effective protection and restoration of much of North America's rich biological heritage and be able to pass on real forests to future generations.

First Postscript

This book and the experiences recounted herein represent early steps in an intensifying struggle to protect temperate forest biodiversity through a combination of conservation biology, law, and policy. Thus, the events described below are but an initial postscript, with others to be written.

As this book was in press, the first federal court rulings in the Wisconsin cases were decided. The court held that during the planning period under review (the early 1980s), the application of conservation biology was still too uncertain to require analysis and disclosure in the Environmental Impact Statements accompanying the Chequamegon and Nicolet National Forest Plans. These decisions suggest that the Wisconsin cases were before their time. As the science described in this book becomes increasingly acknowledged as fundamental to understanding the fate of ecosystems, the one-time exemption granted in this first court decision is unlikely to be accepted as a valid basis for ignoring relevant science in forest management. The judge's decisions left for another day the deeper questions of how the diversity mandate in federal law should be addressed (Chapter 13).

As this book goes to production, appeals are pending which criticize the trial court's decision to dismiss evolving scientific expertise due to inevitable uncertainties in application. Our appeals, and future litigation, will need to address the continuing scientific uncertainty surrounding losses in biodiversity. This uncertainty, and its corollary—the need for cautious management—formed central themes for this book. Developing policy in the face of considerable uncertainty also represents a central problem for other "strategic" environmental problems that involve long-term predictions of environmental harm, including global warming and ozone depletion.

If the courts remain unwilling to compel application of science's long-term predictive capabilities, we may need to insist that our domestic law and policy regarding biodiversity incorporate the "precautionary principle" increasingly utilized in international environmental law (Hey 1992; Van Dyke 1993). We may need to shift the burden of proof to require that proof of harmlessness be furnished by those who propose the intensive disturbance of the landscape and claim to have the wisdom to fully understand the implications of that use.

References

AFSEEE (Association of Forest Service Employees for Environmental Ethics). 1993. What the Tongass didn't want to know—Saga of the V-Pop Report. *Inner Voice* 5(3): 6–7.

Ahlgren, C. E. 1974. Effects of fires on temperate forests: North Central United States. Pages 195–223 in T.T. Kozlowski and C.E. Ahlgren (eds.), *Fire and Ecosystems*. New York: Academic Press.

Ahlgren, C.E. and I. Ahlgren. 1984. *Lob Trees in the Wilderness*. Minneapolis, Minnesota: University of Minnesota Press.

Albert, D. A. 1993. Draft ecoregional map and classification of Michigan, Minnesota, and Wisconsin. Lansing: Michigan Natural Features Inventory.

Allen, D. 1979. *The Wolves of Minong: Their Vital Role in Wild Communities*. Boston: Houghton-Mifflin.

Allen, E.B. and L.L. Jackson. 1992. The arid West. *Restoration and Management Notes* 10(1): 56–59.

Allen, S.W. and G.W. Sharpe. 1960. *An Introduction to American Forestry*. New York: McGraw-Hill Book Company.

Alverson, W.S. and D.M. Waller. 1993. Is it un-biocentric to manage? *Wild Earth* 2(4): 9–10.

Alverson, W.S., D.M. Waller and S.L. Solheim. 1988. Forests too deer: Edge effects in northern Wisconsin. *Conservation Biology* 2: 348–358. (See also Erratum in *Conservation Biology* 3(1): 9, 1989.)

Ambuel, B. and S.A. Temple. 1983. Area-dependent changes in the bird communities and vegetation of southern Wisconsin forests. *Ecology* 64: 1057–1068.

Andersen, R.A. 1992. Diversity of eukaryotic algae. *Biodiversity and Conservation* 1(4): 267–292.

Andrén, H. and P. Angelstam. 1988. Elevated predation rates as an edge effect in habitat islands: Experimental evidence. *Ecology* 69: 544–547.

Andrén, H., P. Angelstam, E. Linström and P. Widén. 1985. Differences in predation pressure in relation to habitat fragmentation: An experiment. *Oikos* 45: 273–277.

Bader, M. 1992. A Northern Rockies Proposal for Congress. *Wild Earth*. Special Issue: "The Wildlands Project: Plotting a North American Wilderness Recovery Strategy." Canton, New York: Cenozoic Society, Inc.

Barnes, B.V. 1989. Old growth forests of the northern Lake States: A landscape ecosystem perspective. *Natural Areas Journal* 9: 45–57.

Barney, D.R. 1974. *The Last Stand: Ralph Nader's Study Group Report on the National Forests*. New York: Grossman Publishers.

Basler, B. 1993. Vietnam forest yields evidence of new animal. *The New York Times*, June 8, 1993, pp. B7, B9.

Beard, J. 1991. Woodland soil yields a multitude of insect species. *New Scientist* 131: 18.

Beier, P. 1993. Determining minimum habitat areas and habitat corridors for Cougars. *Conservation Biology* 7: 94–108.

Bellinger, R.G., F.W. Ravlin and M.L. McManus. 1989. Forest edge effects and their influence on gypsy moth (Lepidoptera: Lymantriidae) egg mass distributions. *Environmental Entomology* 18(5): 840–843.

Benjamin, S.E. and L.A. Maguire. 1991. Effects of road edges of the herb communities of fragmented piedmont forests (abstract). Abstracts, Fifth Annual Meeting of the Society for Conservation Biology, Madison, Wisconsin, p. 82.

Bennett, A.F. 1991. Roads, roadsides and wildlife conservation: A review. Pages 99–117 in D.A. Saunders and R.J. Hobbs (eds.), *Nature Conservation 2: The Role of Corridors*. Chipping Norton, Australia: Surrey Beatty and Sons.

Blake, J.G. 1991. Nested subsets and the distribution of birds on isolated woodlots. *Conservation Biology* 5: 58–66.

Blockstein, D.E. and H.B. Tordoff. 1985. Gone forever: A contemporary look at the extinction of the passenger pigeon. *American Birds* 39: 845–851.

Blouch, R.I. 1984. [Deer in the] Northern Great Lakes States and Ontario forests. Pages 391–410 in L.K. Halls (ed.), *The White-tailed Deer: Ecology and Management*. Harrisburg, Pennsylvania: Stackpole Books.

Bock, C. E., J. H. Bock and H. M. Smith, 1993. Proposal for a system of federal livestock exclosures on public rangelands in the western United States. *Conservation Biology* 7(3): 731–733.

Böhning-Gaese, K., M.L. Taper and J.H. Brown. 1993. Are declines in North American insectivorous songbirds due to causes on the breeding range? *Conservation Biology* 7(1): 76–86.

Bolle, A. 1970. A university view of the Forest Service. *Report to the Committee on Interior and Insular Affairs*. U.S. Senate. Washington D.C.

Bolle, A.W., et al. 1993. The scientific burden (scientist's letter to Vice-President Gore). *Inner View* 5(3):4

Bormann, F.H. and G.E. Likens. 1979a. Catastrophic disturbance and the steady state in Northern Hardwood forests. *American Scientist* 67: 660–669.

Bormann, F.H. and G.E. Likens. 1979b. *Pattern and Process in a Forested Ecosystem*. New York: Springer-Verlag.

Bormann, F.H., G.E. Likens, D.W. Fisher and R.S. Pierce. 1968. Nutrient loss accelerated by clearcutting of a forest ecosystem. *Science* 159: 882–884.

Bormann, F.H., G.E. Likens, T.G. Siccama, R.S. Pierce and J.S. Eaton. 1974. The export of nutrients and recovery of stable conditions following deforestation at Hubbard Brook. *Ecological Monographs* 44: 255–277.

Botkin, D. 1990. *Discordant Harmonies: A New Ecology for the Twenty-First Century*. New York: Oxford University Press.

Botkin, D.B., J.F. Janak and J.R. Wallis. 1972. Some ecological consequences of a computer model of forest growth. *Journal of Ecology* 60: 849–873.

Bowles, J.B. 1963. *Ornithology of changing forest stands on the western slope of the Cascade Mountains in Central Washington*. M.S. Thesis. University of Washington, Seattle.

Bowyer, J.L. 1992. Responsible Environmentalism: Forests, People, Raw Material Needs, and Environmental Protection. Hamilton Roddis Memorial Lecture, 22 October 1992. Department of Forestry. University of Wisconsin—Madison.

Bradshaw, F.J. 1992. Quantifying edge effect and patch size for multiple-use silviculture: A discussion paper. *Forest Ecology and Management* 48(3/4): 249–264.

Bratton, S.P. 1975. The effect of European wild boar (*Sus scrofa*) on gray beech forest in the Great Smoky Mountains. *Ecology* 56: 1356–1366.

Bratton, S.P. 1979. Impacts of white-tailed deer on the vegetation of Cades Cove, Great Smoky Mountains National Park. *Proceedings of the Annual Conference of the Southeastern Association of Fish and Wildlife Agencies* 33: 305–312.

Bratton, S.P. 1994. Logging and fragmentation of broadleaved deciduous forests: Asking the right ecological questions. *Conservation Biology* 8(1): 295–297.

Braun, E.L. 1950. *Deciduous Forests of Eastern North America.* Philadelphia: Blakiston Company.

Brenneman, R.E. and T.R. Eubanks. 1988. Forest fragmentation in the Northeast—An industry perspective. Pages 23–26 in R.M. DeGraaf and W.M. Healy (eds.), *Is Forest Fragmentation a Management Issue in the Northeast?* Rochester, New York: USDA Forest Service Northeastern Forest Experiment Station.

Brewer, R. 1980. A half-century of changes in the herb layer of a climax deciduous forest in Michigan. *Journal of Ecology* 68: 823–832.

Bridgewater, P.B. 1987. Connectivity: An Australian perspective. Pages 195–200 in A. Saunders, G.W. Arnold, A.A. Burbidge and A.J.M. Hopkins (eds.), *Nature Conservation: The Role of Remnants of Native Vegetation.* Chipping Norton, Australia: Surrey Norton & Sons Pty Ltd.

Britell, J. 1992. The failure of Shasta Costa: How the Agency's New Perspective flagship ran aground. *Inner Voice* 4(4), pp. 9, 14, 15.

Brittingham, M.C. and S.A. Temple. 1983. Have cowbirds caused forest songbirds to decline? *BioScience* 33: 31–35.

Broad, W. J. 1993a. A voyage into the abyss: Gloom, Gold and Godzilla. *The New York Times,* November 2, pp. B5, B8.

Broad, W. J. 1993b. Strange new microbes hint at a vast subterranean world. *The New York Times,* December 28, pp. B5, B8.

Brocke, R.H., J.P. O'Pezio and K.A. Gustafson. 1988. *A forest management scheme mitigating impact of road networks on sensitive wildlife species.* Society of American Foresters. Rochester, New York: USDA Forest Service Northeastern Forest Experiment Station.

Brothers, T.S. 1993. Fragmentation and edge effects in central Indiana old-growth forests. *Natural Areas Journal* 13(4): 268–275.

Brown, C.D. 1993. Mapping old growth. *Audubon* [May/June]: 131–133.

Brown, R.N. and R.S. Lane. 1992. Lyme disease transmission in California: A novel enzootic transmission cycle of *Borrelia burgdorferi*. *Science* 256: 1439–1442.

Brussard, J.M., D.D. Murphy and R.F. Noss. 1992. Strategy and tactics for conserving biological diversity in the United States. *Conservation Biology* 6(2): 157–159.

Bryant, J. 1993. Forest Biodiversity and Clearcutting Prohibition Act of 1993. H.R. 1164, U.S. Congress.

Bull, A.T. 1992. Microbial diversity. *Biodiversity and Conservation* 1(4): 219–220.

Burger, G.V. 1973. *Practical Wildlife Management.* New York: Winchester Press.

Burgess, R.L. and D.M. Sharpe, Eds. 1981. *Forest Island Dynamics in Man-dominated Landscapes.* New York: Springer-Verlag.

Burkman, W., Q. Chavez, R. Cooke, S. Cox, S. DeLost, T. Luther, M. Mielke, M. Miller-Weeks, F. Peterson, M. Roberts, P. Sever and D. Twardus. 1993. *Northeastern Area Forest Health Report.* USDA Forest Service, Northeastern Area report NA-TP-03–93.

Cairns, J. 1986. The myth of the most sensitive species. *BioScience* 36: 670–672.

Camarena, A. [Acting Forest Supervisor, Siskiyou National Forest] and staff. 1990. *Shasta Costa from a New Perspective.* USDA Forest Service publication 794–167.

Campbell, D.G. and H.G. Hammond. 1989. *Floristic Inventory of Tropical Countries.* Bronx, New York: The New York Botanical Garden.

Campbell, F. T. and S. E. Schlarbaum. 1994. *Fading Forests: North American Trees and the Threat of Exotic Pests.* Washington, D.C.: Natural Resources Defense Council.

Canham, C. D. 1978. Catastrophic windthrow in the Hemlock–Harwood forest of Wisconsin. MS. Thesis, Department of Botany, University of Wisconsin, Madison.

Canham, C.D. and O.L. Loucks. 1984. Catastrophic windthrow of the presettlement forests of Wisconsin. *Ecology* 65: 803–809.

Carey, A. B. 1989. Wildlife associated with old-growth forests in the Pacific Northwest. *Natural Areas Journal* 9(3): 151–162.

Carlquist, S. 1974. *Island Biology.* New York: Columbia University Press.

Cenozoic Society. 1992. The Wildlands Project: Plotting a North American Wilderness Recovery Strategy. *Wild Earth*, special issue.

Chabrek, R.H. and A.W. Palmisano. 1973. The effects of Hurricane Camille on the marshes of the Mississippi River delta. *Ecology* 54: 1118–1123.

Chen, J., J.F. Franklin and T.A. Spies. 1992. Vegetation responses to edge environments in old-growth Douglas-fir forests. *Ecological Applications* 2(4): 387–396.

Clary, D.A. 1986. *Timber and the Forest Service.* Lawrence, Kansas: University of Kansas Press.

Clawson, M. 1975. *Forests for Whom and for What?* Baltimore, Maryland: The Johns Hopkins University Press.

Cleland, D.T., T.R. Crow, P.E. Avers, and J.R. Probst. 1992. Principles of land stratification for delineation ecosystems. Presented at "Taking and Ecological Approach to Management" National Workshop, USDA Forest Service, Washington, D.C.

Clements, F.E. 1916. *Plant Succession: An Analysis of the Development of Vegetation.* Carnegie Institution of Washington Publication 242.

Cody, M.L. 1975. Towards a theory of continental species diversities: Bird distributions over Mediterranean habitat gradients. Pages 214–257 in M.L. Cody and J.M. Diamond (eds.), *Ecology and Evolution of Communities.* Cambridge, Massachusetts: Harvard University Press.

Cody, M.L. 1986. Diversity, rarity, and conservation in Mediterranean-climate regions. Pages 122–152 in M.E. Soulé (ed.), *Conservation Biology: The Science of Scarcity and Diversity.* Sunderland, Massachusetts: Sinauer Associates.

Coffman, M.S. 1992. *Environmentalism! The Dawn of Aquarius or the Twilight of a New Dark Age?* Bangor, Maine: Environmental Perspectives.

Colinvaux, P. 1986. *Ecology.* New York: J. Wiley & Sons.

Concannon, J.A., P.B. Alaback and J.F. Franklin. 1990. Edge effects of a forest island in southeast Alaska (abstract). *Northwest Science* 64(2): 94.

Connell, J.H. 1978. Diversity in tropical rain forests and coral reefs. *Science* 199: 1302–1310.

Connell, J.H. and R.O. Slatyer. 1977. Mechanisms of succession in natural communities and their role in community stability and organization. *American Naturalist* 111: 1119–1144.

Cooley, J.L. and J.H. Cooley, Eds. 1984. *Natural Diversity in Forest Ecosystems.* Proceedings of a workshop held November 29–December 1, 1982. Institute of Ecology, University of Georgia, Athens.

Cromack, K., Jr., B.L. Fichter, A.M. Moldenke, J.A. Entry and E.R. Ingham. 1988. Interactions between soil animals and ectomycorrhizal fungal mats. *Agriculture, Ecosystems and Environment* 24: 161–168.

Cronon, W. 1983. *Changes in the Land: Indians, Colonists, and the Ecology of New England.* New York: Hill and Wang.

Cronon, W. 1991. *Nature's Metropolis: Chicago and the Great West.* New York: W.W. Norton.

Crow, J.F. and M. Kimura. 1970. *An Introduction to Population Genetics Theory.* Minneapolis, Minnesota: Burgess Publishing Company.

Crow, T.R., A. Haney and D.M. Waller. 1993. Report of the Scientific Roundtable on Biological Diversity. *USDA Forest Service publication TP-R9-CHF/NNF-93-1.* Washington, D.C.: U.S. Government Printing Office.

Crumpacker, D.W., S.W. Hodge, D. Friedley and W.P. Gregg. 1988. A preliminary assessment of the status of major terrestrial and wetland ecosystems on federal and Indian lands in the U.S. *Conservation Biology* 2: 103–115.

Csuti, B. 1991. Introduction [to part 2, Conservation Corridors: Countering Habitat Fragmentation]. Pages 81–90 in W.E. Hudson (ed.), *Lanscape Linkages and Biodiversity.* Washington, D.C.: Island Press.

Curtis, J.T. 1959. *The Vegetation of Wisconsin.* Madison: University of Wisconsin Press.

Cutler, A. 1991. Nested faunas and extinction in fragmented habitats. *Conservation Biology* 5: 496–505.

Dahlberg, B.L. and R.C. Guettinger. 1956 *The White-tailed Deer in Wisconsin.* Madison: Wisconsin Conservation Department.

Darlington, P.J. 1957. *Zoogeography: The Geographical Distribution of Animals.* New York: J. Wiley & Sons.

Dasmann, R.F. 1964. *Wildlife Biology.* New York: J. Wiley & Sons.

Davis, M.B. 1965. Phytogeography and palynology of northeastern United States. Pages 377–401 in H.E. Wright and D.G. Frey (eds.), *The Quaternary of the United States*. Princeton, New Jersey: Princeton University Press.

Davis, M.B. 1969. Palynology and environmental history during the Quaternary Period. *American Scientist* 57: 317–332.

Davis, M.B. 1993. *Old Growth in the East: A Survey*. Richmond, Vermont: Cenozoic Society, Inc.

DeBell, D.S. 1990. *Sustainable Forestry: Social, Economic, and Ecological Considerations*. Washington, D.C.: Society of American Foresters.

deCalesta, D. 1992. Impact of deer on diversity of plants, songbirds, and small mammals (abstract). In *Deer Density Effects on a Forest Ecosystem: A Pennsylvania Story; Results of a 10-Year Study*. Warren, Pennsylvania: Society of American Foresters.

deCalesta, D.S. 1994a. Impact of deer on interior forest songbirds in northwestern Pennsylvania. *Journal of Wildlife Management*, in press.

deCalesta, D.S. 1994b. Impact of deer and silvicultural treatments on small mammals in northwestern Pennsylvania. *Journal of Mammology*, in press.

DeGraaf, R.M. 1992. Effects of even-aged management on forest birds at northern hardwood stand interfaces. *Forest Ecology and Management* 47(1–4): 95–110.

Delcourt, H.R., P.A. Delcourt and T. Webb. 1983. Dynamic plant ecology: The spectrum of vegetational change in space and time. *Quaternary Science Review* 1: 153–175.

Delcourt, H.R. and P.A. Delcourt. 1991. *Quaternary Ecology: A Paleoedological Perspective*. London: Chapman & Hall.

Delcourt, P.A. and H.R Delcourt. 1992. Ecotone dynamics in space and time. Pages 19–54 in A.J. Hansen and F. DiCastri, *Landscape Boundaries: Consequences for Biotec Diversity and Ecological Flows*. New York: Springer-Verlag.

Denevan, W.M., Ed. 1992. *The Native Population of the Americas in 1492*. Madison: University of Wisconsin Press.

Denslow, J.S. 1985. Disturbance mediated coexistence of species. Pages 307–323 in S.T.A. Pickett and P.S. White (eds.), *The Ecology of Natural Disturbance and Patch Dynamics*. New York: Academic Press.

Devall, B. 1993. *Clearcut: The Tragedy of Industrial Forestry*. San Francisco: Sierra Club Books and Earth Island Press.

Devall, M.S. and P.F. Ramp. 1992. U.S. Forest Service Research Natural Areas and protection of old growth in the South. *Natural Areas Journal* 12(2): 75–85.

Diamond, J.M. 1972. Biogeographic kinetics: Estimation of relaxation times for avifaunas of Southwest Pacific islands. *Proceedings of the National Academy of Sciences, USA* 69: 3199–3203.

Diamond, J.M. 1973. Distributional ecology of New Guinea birds. *Science* 179: 759–769.

Diamond, J.M. 1975. Assembly of species communities. Pages 342–444 in M.L. Cody and J.M. Diamond (eds.), *Ecology and Evolution of Communities*. Cambridge, Massachusetts: Harvard University Press.

Diamond, J.M. 1976. Island biogeography and conservation: Strategy and limitations. *Science* 193: 1027–1032.

Diamond, J.M. 1985a. How many unknown species are yet to be discovered? *Nature* 315: 538–539.

Diamond, J.M. 1985b. In quest of the wild and weird. *Discover* 3/85: 34–41.

Diamond, J.M. 1990. The search for life on Earth. *Natural History* 4/90: 28–32.

Diamond, J.M. 1992. Must we shoot deer to save nature? *Natural History* 8/92: 2–8.

Disney, R.H.L. 1989. Does anyone care? *Conservation Biology* 3(4): 414.

DNR (Department of Natural Resources, State of Wisconsin). 1989. *Wisconsin Timber Wolf Recovery Plan.* Wisconsin Endangered Resources Report 50. Madison: Wisconsin.

DNR (Department of Natural Resources, State of Wisconsin). 1992. *The 1991 Wisconsin Deer Harvest Report.* Madison: Wisconsin.

Dobson, A.P. and R.M. May. 1986. Disease and conservation. Pages 534–365 in M.E. Soulé (ed.), *Conservation Biology: The Science of Scarcity and Diversity.* Sunderland, Massachusetts: Sinauer Associates.

Duffus, D. 1993. Tsitika to Baram: The myth of sustainability. *Conservation Biology* 7(2): 440–442.

Duffy, D.C. 1993a. Herbs and clearcutting: Reply to Elliot and Loftis and Steinbeck. *Conservation Biology* 7(2): 221–222.

Duffy, D.C. 1993b. Seeing the forest for the trees: Response to Johnson et al. *Conservation Biology* 7(2): 436–439.

Duffy, D.C. and A.J. Meier. 1992. Do Appalachian herbaceous understories ever recover from clearcutting? *Conservation Biology* 6: 196–201.

Egan, T. 1993. Upheaval in the forests—Clinton plan shifts emphasis from logging but does not create off-limits wilderness. *The New York Times,* July 2, 1993, pp. A1, A9.

Ehrenfeld, D.W. 1978. *The Arrogance of Humanism.* New York: Oxford University Press.

Ehrenfeld, D.W. 1989. Is anyone listening? *Conservation Biology* 3(4): 415.

Ehrenfeld, D.W. 1993. *Beginning Again: People and Nature in the New Millennium.* New York: Oxford University Press.

Ehrlich, A.H. and P.R. Ehrlich. 1980. The Greeks and Romans did it, too! *The Mother Earth News* (May/June): 148–149.

Ehrlich, P.R. 1988. The loss of diversity: Causes and consequences. Chapter 2, pages 21–27 in E.O. Wilson (ed.), *Biodiversity.* Washington, D.C.: National Academy Press.

Ehrlich, P.R. and A.H. Ehrlich. 1981. *Extinction: The Causes and Consequences of the Disappearance of Species.* New York: Random House.

Ehrlich, P.R. and A.H. Ehrlich. 1991. *Healing the Planet: Strategies for Resolving the Environmental Crisis.* Reading, Massachusetts: Addison-Wesley.

Ehrlich, P.R. and E.O. Wilson. 1991. Biodiversity studies: Science and policy. *Science* 253: 758–762.

Eldredge, N. 1992. *Systematics, Ecology and the Biodiversity Crisis.* New York: Columbia University Press.

Elliott, D.K., Ed. 1986. *Dynamics of Extinction.* New York: John Wiley & Sons.

Elliott, K.J. and D.L. Loftis. 1993. Vegetation diversity after logging in the southern Appalachians. *Conservation Biology* 7(2): 220–221.

Elton, C.S. 1958. *The Ecology of Invasion by Animals and Plants.* London: Methuen.

Ervin, K. 1989. *Fragile Majesty: The Battle for North America's Last Great Forest.* Seattle, Washington: The Mountaineers.

Evans, H. 1990. Public affairs. *Nicolet News* 122. Rhinelander, Wisconsin: Nicolet National Forest.

Ewald, P.W. 1988. Cultural vectors, virulence, and the emergence of evolutionary epidemiology. *Oxford Surveys in Evolutionary Biology* 5: 216.

Fahey, T.J. and W.A. Reiners. 1981. Fire in the forests of Maine and New Hampshire. *Bulletin of the Torrey Botanical Club* 108: 362–373.

Falk, D.A. and K.E. Holsinger, Eds. 1991. *Genetics and Conservation of Rare Plants.* New York: Oxford University Press.

Farnum, P., R. Timmis and J.L. Kulp. 1983. Biotechnology of forest yield. *Science* 219: 694–702.

Ferry, B.W., M.S. Baddeley and D.L. Hawksworth, Eds. 1973. *Air Pollution and Lichens.* London: Athlone Press.

Findley, R. 1990. Will we save our own? *National Geographic* 178(3): 106–136.

Foreman, D. 1991a. *Confessions of an Eco-Warrior.* New York: Harmony Books.

Foreman, D. 1991b. The new conservation movement. *Wild Earth* 2: 6-12.

Foreman, D. 1992. Developing a regional wilderness recovery plan. *Wild Earth* (The Wildlands Project Special Issue): 26–29.

Forman, R.T.T. 1974. Ecosystem size: A stress on the forest ecosystem. *Bulletin of the Ecological Society of America* 55(2): 38.

Forman, R.T.T. and M. Godron. 1986. *Landscape Ecology.* New York: John Wiley & Sons.

Forzley, K.C., T.A. Grudzien and J.R. Wells. 1993. Comparative floristics of seven islands in northwestern Lake Michigan. *Michigan Botanist* 32: 3–22.

Fowells, H.A. 1965. *Silvics of Forest Trees of the United States.* Washington, D.C.: U.S. Dept. of Agriculture.

Fox, S.R. 1981. *The American Conservation Movement: John Muir and His Legacy.* Madison: University of Wisconsin Press.

Franklin, J. 1989. Toward a new forestry. *American Forests* 95 (Nov./Dec.): 37–44.

Franklin, J. 1990. Thoughts on applications of silvicultural systems under New Forestry. *Forest Watch* (Nov./Dec.): 9–11.

Franklin, J.F. 1992. Scientific basis for new perspectives in forests and streams. Pages 25–72 in R.J. Naiman (ed.), *New Perspectives in Watershed Management.* New York: Springer-Verlag.

Franklin, J.F. 1993. Preserving biodiversity: Species, ecosystems, or landscapes? *Ecological Applications* 3(2): 202–205.

Frelich, L.E. 1992. The relationship of natural disturbances to white pine stand development. In R.A. Stine and M.J. Baughman (eds.), *White Pine Symposium Proceedings: History, Ecology, Policy and Management.* Duluth: Minnesota Extension Service, University of Minnesota (NR-BU-6044).

French, J.B., Jr., W.L. Schell, J.J. Kazmierczak and J.P. Davis. 1994. Changes in population density and distribution of *Ixodes dammini* (Acari: Ixodidae) in Wisconsin during the 1980s. *Journal of Medical Entomology*, in press.

Frissell, C.A., R.K. Nawa and R. Noss. 1992. Is there any conservation biology in "New Perspectives?": A response to Salwasser. *Conservation Biology* 6: 461–464.

Frome, M. 1962. *Whose Woods These Are: The Story of the National Forests*. Garden City, New York: Doubleday and Co.

Frome, M. 1974. *Battle for the Wilderness*. New York: Praeger Publishers.

Futuyma, D.J. 1986. *Evolutionary Biology*. Sunderland, Massachusetts: Sinauer Associates.

Garrott R.A., P.J. White and C.A. Vanderbilt White. 1993. Overabundance: An issue for conservation biologists? *Conservation Biology* 7(4): 946–949.

Gaston, K.J. 1991a. The magnitude of global intact species richness. *Conservation Biology* 5(3): 283–297.

Gaston, K.J. 1991b. Estimates of the near-imponderable: A reply to Erwin. *Conservation Biology* 5(4): 564–566.

Gates, D.M., C.H.D. Clarke and J.T. Harris. 1983. Wildlife in a changing environment. Pages 52–80 in S.L. Flader (ed.), *The Great Lakes Forest—An Environmental and Social History*. Minneapolis, Minnesota: University of Minnesota Press.

Gates, J.E. 1991. Powerine corridors, edge effects, and wildlife in forested landscapes of the central Appalachians. Pages 13–32 in J.E. Rodiek and E.G. Bolen (eds.), *Wildlife and Habitats in Managed Landscapes*. Frostburg, Maryland: Appalachian Environmental Lab.

Gates, J.E. and N.R. Giffen. 1991. Neotropical migrant birds and edge effects at a forest-stream ecotone. *Wilson Bulletin* 103(2): 204–217.

Gates, J.E. and L.W. Gysel. 1978. Avian nest dispersion and fledging success in field forest ecotones. *Ecology* 59: 871–883.

Gilbert, F.F. 1974. *Parelaphostrongylus tenuis* in Maine: II. Prevalence in moose. *Journal of Wildlife Management* 38: 42–46.

Gillis, A.M. 1990. The new forestry: An ecosystem approach to land management. *BioScience* 40(8): 558–562.

Gilpin, M. and I. Hanski, Eds. 1991. *Metapopulation Dynamics: Empirical and Theoretical Investigations*. New York: Academic Press.

Givnish, T.J. 1981. Serotiny, geography, and fire in the Pine Barrens of New Jersey. *Evolution* 35: 101–123.

Goldsmith, F.B. 1991. *Monitoring for Conservation and Ecology*. New York: Chapman and Hall.

Graham, E. 1947. *The Land and Wildlife*. New York: Oxford University Press.

Greeley, W.B. 1925. The relation of geography to timber supply. *Economic Geography* 1(1): 1–14.

Grime, J.P. 1979. *Plant Strategies and Vegetation Processes*. New York: J. Wiley & Sons.

Guntenspergen, G. 1983. The minimum size for nature preserves: Evidence from southeastern Wisconsin forests. *Natural Areas Journal* 3(4): 38–46.

Hansen, A., D.L. Urban and B. Marks. 1992. Avian community dynamics: The interplay of landscape trajectories and species life histories. Pages 171–195 in A.J. Hansen and F. Dicastri (eds.), *Landscape Boundaries: Consequences for Biotic Diversity and Ecological Flows*. New York: Springer-Verlag.

Hansen, A.J., T.A. Spies, F.J. Swanson and J.L. Ohmann. 1991. Conserving biodiversity in managed forests. *BioScience* 41: 382–392.

Hansen, E.M. and W. Littke. 1993. *Phytophthora Root Rot and Other Diseases of Yew in Forests and Cultivation*. International Yew Resources Conference: Yew (*Taxus*) conservation biology and interactions. Berkeley, California: USDA Forest Service.

Harmon, M.E., W.K. Ferrell and J.F. Franklin. 1990. Effects of carbon storage of conversion of old-growth forests to young forests. *Science* 247: 699–702.

Harper, J. 1977. *Population Biology of Plants*. New York: Academic Press.

Harris, L.D. 1984. *The Fragmented Forest: Island Biogeography Theory and the Preservation of Biotic Diversity*. Chicago: The University of Chicago Press.

Harris, L.D. 1988a. Edge effects and conservation of biotic diversity. *Conservation Biology* 2(4): 330–332.

Harris, L.D. 1988b. Reconsideration of the habitat concept. Pages 137–144 in R.E. McCabe (ed.), *Transactions of the Fifty-third North American Wildlife and Natural Resources Conference*. Washington, D.C.: Wildlife Management Institute.

Harris, L.D. and K. Atkins. 1991. Faunal movement corridors in Florida. Pages 117–134 in W.E. Hudson (ed.), *Landscape Linkages and Biodiversity*. Washington, D.C.: Island Press.

Harris, L.D. and P.B. Gallagher. 1989. New initiatives for wildlife conservation: The need for movement corridors. Pages 11–34 in G. Mackintosh (ed.), *In Defense of Wildlife: Preserving Communities and Corridors*. Washington, D.C.: Defenders of Wildlife.

Harris, L.D. and J. Scheck. 1991. From implications to applications: The dispersal corridor principle applied to the conservation of biological diversity. Pages 189–220 in D.A. Saunders and R.J. Hobbs (eds.), *Nature Conservation 2: The Role of Corridors*. Chipping Norton, Australia: Surrey Beatty & Sons Pty Ltd.

Harris, L.D., C. Maser and A. McKee. 1982. Patterns of old growth harvest and implications for Cascades wildlife. *Transactions of the North American Wildlife Natural Resources Conference* 47: 374–392.

Heinselman, M.L. 1973. Fire in the virgin forests of the Boundary Waters Canoe Area, Minnesota. *Quaternary Research* 3: 329–382.

Heinselman, M.L. 1981. Fire and succession in the conifer forests of northern North America. Pages 374–405 in D.C. West, II.II. Shugart and D.B. Botkin (eds.), *Forest Succession: Concepts and Application*. New York: Springer-Verlag.

Henry, J.D. and J.M.A. Swan. 1974. Reconstructing forest history from live and dead plant material—An approach to the study of forest succession in southwest New Hampshire. *Ecology* 55: 772–783.

Hey, E. 1992. The precautionary concept in environmental policy and law: Institutionalizing caution. *The Georgetown International Environmental Law Review* 4: 303–318.

Hickman, S. 1990. Evidence of edge species attraction to nature trails within deciduous forest. *Natural Areas Journal* 10(1): 3–5.

High Country News (staff). 1991. Two say politics rule their agencies. *High Country News* 23(18): 1-10 (10/7/91).

Hilsenhoff, W.L. 1982. Using a biotic index to evaluate water quality in streams. *Wisconsin Department of Natural Resources Technical Bulletin No. 132.*

Hirt, P.W. 1991. *Sustained Yield, Multiple Use, and Intensive Management on the National Forests, 1945–1991.* Ph.D. Dissertation. University of Arizona.

Hobbs, R.J. 1992. The role of corridors in conservation: Solution or bandwagon? *Trends in Ecology and Evolution* 7(11): 389–392.

Hobbs, R.J. and A.J.M. Hopkins. 1991. The role of conservation corridors in a changing climate. Pages 281–290 in D.A. Saunders and R.J. Hobbs (eds.), *Nature Conservation 2: The Role of Corridors.* Chipping Norton, Australia: Surrey Beatty & Sons Pty Ltd.

Hodkinson, I.D. 1992. Global insect diversity revisited. *Journal of Tropical Ecology* 8: 505–508.

Hoppe, D.J. 1990. Nicolet National Forest ecological classification and inventory proposal information, February 11, 1990. Rhinelander, Wisconsin: unpublished document. Nicolet National Forest.

Horn, H.S. 1971. *The Adaptive Geometry of Trees.* Princeton, New Jersey: Princeton University Press.

Horn, H.S. 1975. Forest succession. *Scientific American* 232(5): 91–98.

Horn, H.S. 1981. Some causes of variety in patterns of secondary succession. Pages 24–35 in D.C. West, H.H. Shugart and D.B. Botkin (eds.), *Forest Succession: Concepts and Application.* New York: Springer-Verlag.

Horsley, S.B. 1992. Impact of deer on interactions among plants (abstract). In *Deer Density Effects on a Forest Ecosystem: A Pennsylvania Story; Results of a 10-Year Study.* Warren, Pennsylvania: Society of American Foresters.

Horsley, S.B. 1993. Mechanisms of interference between hayscented fern and black cherry. *Canadian Journal of Forestry Research* 23: 2059—2069.

Hough, A.F. 1936. A climax forest community on East Tionesta Creek in northwestern Pennsylvania. *Ecology* 17(1): 9–28.

Hough, A.F. 1965. A twenty-year record of understory vegetational change in a virgin Pennsylvania forest. *Ecology* 46(3): 370–373.

Hough, A.F. and R.D. Forbes. 1943. The ecology and silvics of forests in the high plateaus of Pennsylvania. *Ecological Monographs* 13: 299–320.

Houghton, R.A. and G.W. Woodwell 1989. Global climate change. *Scientific American* 260(April): 36–44.

Howe, R.W., S.A. Temple and M.J. Mossman. 1992. Forest management and birds in northern Wisconsin. *Passenger Pigeon* 54: 297–305.

Hunter, M.L., Jr. 1989. What constitutes an old-growth stand? *Journal of Forestry* 87(8): 33–35.

Hunter, M.L., Jr. 1990. *Wildlife, Forests, and Forestry: Principles of Managing Forests for Biological Diversity.* Englewood Cliffs, New Jersey: Prentice Hall.

Hunter, M.L., Jr. 1991. Towards a conceptual definition of old-growth forests (abstract). p. 78, in *Abstracts, Fifth Annual Meeting of the Society for Conservation*

Biology, Madison, Wisconsin. Cambridge, Massachusetts: Blackwell Scientific Publications.

Hunter, M.L., Jr., G.L. Jacobson, Jr. and T. Webb III. 1988. Paleoecology and the coarse-filter approach to maintaining biological diversity. *Conservation Biology* 2(4): 375–385.

Iltis, H. H. 1970. Man first? Man last? The paradox of human ecology. *BioScience* 20(14): 820.

Irwin, T.L. 1982. Tropical forests: Their richness in coleoptera and other arthropod species. *Coleopterists' Bulletin* 36(1): 74–75.

Irwin, T.L. 1988. The tropical forest canopy, the heart of biotic diversity. Chapter 13, pages 123–129 in E.O. Wilson (ed.), *Biodiversity*. Washington, D.C.: National Academy Press.

Irwin, T.L. 1991. How many species are there?: Revisited. *Conservation Biology* 5(3): 330–333.

Jaenike, J. 1991. Mass extinction of European fungi. *Trends in Ecology and Evolution* 6(6): 174–175.

Janzen, D.H. 1983. No park is an island: Increased interference from the outside as park size decreases. *Oikos* 41: 402–410.

Janzen, D.H. 1986. The eternal external threat. Pages 286–303 in M.E. Soulé (ed.), *Conservation Biology: The Science of Scarcity and Diversity.* Sunderland, Massachusetts: Sinauer Associates.

Janzen, D.H. 1988. Tropical ecological and biocultural restoration. *Science* 239: 243–244.

Johnson, A.S., W.M. Ford and P.E. Hale. 1993. The effects of clearcutting on herbaceous understories are still not fully known. *Conservation Biology* 7(2): 433–435.

Jonkel, C.J. and I.M. Cowan 1971. The black bear in spruce-fir forest. *Wildlife Monographs 27.*

Juday, G.P. 1988. Old growth forests and natural areas: An introduction. *Natural Areas Journal* 8: 3–6.

Judziewicz, E.J. and R.G. Koch. 1993. Flora and vegetation of the Apostle Islands National Lakeshore and Madeline Island, Ashland and Bayfield Counties, Wisconsin. *Michigan Botanist* 32(2): 43–189.

Karamanski, T.J. 1989. *Deep Woods Frontier: A History of Logging in Northern Michigan.* Detroit: Wayne State University Press.

Keene, R. 1993. Salvage logging: Health or hoax? *Inner Voice* 5: 1–4.

Keiter, R.B. and M.S. Boyce, Eds. 1991. *The Greater Yellowstone Ecosystem.* New Haven, Connecticut: Yale University Press.

Kick, D.D. and staff. 1991. *Draft Environmental Impact Statement, Sunken Camp Area, Washburn Ranger District, Chequamegon National Forest.* Park Falls, Wisconsin: USDA Forest Service.

Kirkland, A. 1988. The rise and fall of multiple-use forest management in NZ [*sic*]. *New Zealand Forestry* (May): 9–12.

Klopatek, C.C., E.G. O'Neill, D.W. Freckman, C.S. Bledsoe, D.C. Coleman, D.A. Crossley, Jr., E.R. Ingham, D. Parkinson and J.M. Klopatek, 1992. The sustainable biosphere initiative: A commentary from the U.S. Soil Ecological Society. *Bulletin of the Ecological Society of America* 73(4): 223–228.

Knize, P. 1991. The mismanagement of the National Forests. *The Atlantic Monthly* 268: 98-112.

Kohm, K.A., Ed. 1991. *Balancing on the Brink of Extinction.* Washington, D.C.: Island Press.

Kotar, J., J.A. Kovach and C.T. Locey. 1988. *Field Guide to Forest Habitat Types of Northern Wisconsin.* Madison: Department of Forestry, University of Wisconsin—Madison and Wisconsin Department of Natural Resources.

Krebs, C.J. 1985. *Ecology: The Experimental Analysis of Distribution and Abundance.* New York: Harper and Row.

Kremen, C., R.K. Colwell, T.L. Erwin, D.D. Murphy, R.F. Noss and M.A. Sanjayan. 1993. Terrestrial arthropod assemblages: Their use in conservation planning. *Conservation Biology* 7(4): 769–808.

Kricher, J.C. 1990. The double edge effect. *Bird Observer* 18(2): 80–84.

Kroodsma, R.L. 1984. Ecological effects associated with degree of edge effect in breeding birds. *Journal of Wildlife Management* 48(2): 418–425.

Krugman, S.L. 1990. One person's view of forest management on public lands (review of Robinson's *Excellent Forestry*). *Conservation Biology* 4(2): 205–206.

La Duke, J.C. 1987. The existence of hypotheses in plant systematics, or biting the hand that feeds you. *Taxon* 36(1): 60–64.

Landres, P.B., J. Verner and J.W. Thomas. 1988. Ecologial uses of vertebrate indicator species: A critique. *Conservation Biology* 2: 316–328.

Lansky, M. 1992. *Beyond the Beauty Strip: Saving What's Left of Our Forests.* Gardiner, Maine: Tilbury House.

Lattin, J.D. 1990. Arthropod diversity in northwest old-growth forests. *Wings* (summer): 7–10.

Laurance, W.F. 1991. Edge effects in tropical forest fragments: Application of a model for the design of nature reserves. *Biological Conservation* 57(2): 205–220.

Lawrence, N. and D. Murphy. 1992. New perspectives or old priorities? *Conservation Biology* 6: 465–468.

Legislative Audit Bureau, State of Wisconsin. 1992. *An Evaluation of Deer Management Policies, Department of Natural Resources.* Madison: Legislative Audit Bureau.

Lennartz, M. and R. Lancia. 1989. Old-growth wildlife in second-growth forests: Opportunities for creative silviculture. Pages 74–103 in U.S. Forest Service Timber Management (ed.), *Silviculture for All Resources.* Washington, D.C.: USDA Forest Service.

Leopold, A. 1921. Wilderness and its place in forest recreational policy. *Journal of Forestry* 19: 718-721. Reprinted in Leopold, Aldo. 1991. *The River of the Mother of God and Other Essays.* Madison: The University of Wisconsin Press.

Leopold, A. 1933. *Game Management.* New York: Charles Scribner's Sons.

Leopold, A. 1936. Deer and Dauerwald in Germany I. History. *Journal of Forestry* 34: 366–375.

Leopold, A. 1939. A biotic view of land. Reprinted in Leopold, Aldo. 1991. *The River of the Mother of God and Other Essays*. Madison: The University of Wisconsin Press.

Leopold, A. 1941. Wilderness as a land laboratory. Reprinted in Leopold, Aldo. 1991. *The River of the Mother of God and Other Essays*. Madison: The University of Wisconsin Press.

Leopold, A. 1942. The last stand. Reprinted in Leopold, Aldo. 1991. *The River of the Mother of God and Other Essays*. Madison: The University of Wisconsin Press.

Leopold, A. 1943. *Wisconsin's Deer Problem*. Madison: Wisconsin Conservation Department.

Leopold, A. 1949. *A Sand County Almanac and Sketches Here and There*. Oxford: Oxford University Press.

Leopold, A. 1953. *Round River*. New York: Oxford University Press.

Leopold, D.J., C. Reschke and D.S. Smith. 1988. Old-growth forests of Adirondack Park, New York. *Natural Areas Journal* 8(3): 166–189.

Lesica, P., B. McCune, S.V. Cooper and W.S. Hong. 1991. Differences in lichen and bryophyte communities between old-growth and managed second-growth forests in the Swan Valley, Montana. *Canadian Journal of Botany* 69: 1745–1755.

Leverett, R.T. 1993. Eastern old-growth forest: A new perspective. Forword, pages 8–15 in M.B. Davis, *Old Growth in the East: A Survey*. Richmond, Vermont: Cenozoic Society, Inc.

Levin, S.A. and R.T. Paine. 1974. Disturbance, patch formation, and community structure. *Proceedings of the National Academy of Sciences, U.S.A.* 71: 2744–2747.

Levins, R., T. Awerbuch, U. Brinkmann, I. Eckardt, P. Epstein, N. Makhoul, C. Albuquerque de Passas, C. Puccia, A. Spielman and M.E. Wilson, 1994. The emergence of new diseases. *American Scientist* 82: 52–60.

Linden, E. 1991. Good intentions, woeful results. *Time* (April 1, 1991): 48–49.

Littler, M.M., D.S. Littler, S.M. Blair and J.N. Norris. 1985. Deepest known plant life discovered on an uncharted seamount. *Science* 227: 57–59.

Lorimer, C.G. 1977. The presettlement forest and natural disturbance cycle of northeastern Maine. *Ecology* 58: 139–148.

Lorimer, C.G. 1980. Age structure and disturbance history of a southern Appalachian virgin forest. *Ecology* 61: 1169–1184.

Loucks, O.L. 1970. Evolution of diversity, efficiency, and community stabililty. *American Zoologist* 10: 17–25.

Loucks, O.L. 1992. Forest response research in NAPAP: Potentially successful linkage of policy and science. *Ecological Applications* 2(2): 117–123.

Lovejoy, T.E. 1980. Foreword. Pages ix–x in M.E. Soule and B.A. Wilcox, *Conservation Biology*. Sunderland, Massachusetts: Sinauer Associates.

Lovejoy, T.E. and D.C. Oren. 1981. Minimum critical size of ecosystems. Pages 7–12 in R.L. Burgess and D.M. Sharpe (eds.), *Forest Island Dynamics in Man-Dominated Landscapes*. New York: Springer-Verlag.

Lovejoy, T.E., R.O. Bierregaard, Jr., A.B. Rylands, J.R. Malcolm, C.E. Quintela, L.H. Harper, K.S. Brown, Jr., A.H. Powell, G.V.N. Powell, H.O.R. Schubart and M.B. Hays. 1986. Edge and other effects of isolation on Amazon forest fragments. Pages 257–285 in M.E. Soulé (ed.), *Conservation Biology: The Science of Scarcity and Diversity.* Sunderland, Massachusetts: Sinauer Associates.

Ludwig, D., R. Hilborn and C. Walters. 1993. Uncertainty, resource exploitation, and conservation: Lessons from history. *Science* 260: 17–18.

Luoma, J.R. 1991a. Taxonomy, lacking in prestige, may be nearing a renaissance. *The New York Times,* December 10, 1991, p. B7.

Luoma, J.R. 1991b. A wealth of forest species is found underfoot. *The New York Times,* July 2, 1991, pp. C1, C9.

Luoma, J.R. 1992. Sophisticated tools are giving taxonomy a new lease on life. *The New York Times,* February 18,1992, p. B7.

Lussenhop, J. 1992. Mechanisms of microarthropod-microbial interactions in soil. Pages 1–25 in M. Begon and A.H. Fitter (eds.), *Advances in Ecological Research* 23. London: Academic Press/Harcourt Brace Jovanovich.

Lutz, H. J. 1930. The vegetation of Heart's Content, a virgin forest in northwestern Pennsylvania. *Ecology* 11(1): 1–29.

MacArthur, R.H. 1965. Patterns of species diversity. *Biological Review* 40: 510–533.

MacArthur, R.H. and E.O. Wilson. 1963. An equilibrium theory of insular zoogeography. *Evolution* 17: 373–387.

MacArthur, R.H. and E.O. Wilson. 1967. *The Theory of Island Biogeography.* Princeton, New Jersey: Princeton University Press.

Magurran, A.E. 1988. *Ecology Diversity and Its Measurement.* Princeton, New Jersey: Princeton University Press.

Manes, C. 1990. *Green Rage: Radical Environmentalism and the Unmasking of Civilization.* Boston: Little Brown and Company.

Margulis, L. and K.V. Schwarz. 1987. *Five Kingdoms: An Illustrated Guide to the Phyla of Life on Earth,* second edition. New York: W.H. Freeman.

Marks, P.L. 1974. The role of pin cherry (*Prunus pensylvanica* L.) in the maintenance of stability in northern hardwood ecosystems. *Ecological Monographs* 44: 73–88.

Marquis, D.A. 1981. Effects of deer browsing on timber production in Allegheny hardwood forests in northwestern Pennsylvania. *Research Paper NE-475.* USDA Forest Service, Northeast Forest Experiment Station, Broomall, Pennsylvania.

Marsh, G.P. 1864. *Man and Nature: Or Physical Geography as Modified by Human Action.* New York: Scribners.

Martin, P.S. and R.G. Klein, Eds. 1984. *Quaternary Extinctions: A Prehistoric Revolution.* Tucson, Arizona: University of Arizona Press.

Martin, W. 1992. Characteristics of old-growth mixed mesophytic forests. *Natural Areas Journal* 12(3): 127–135.

Maser, C. 1989. *Forest Primeval.* San Francisco: Sierra Club Books.

Mason, D.J. and T.C. Moermond. 1991. Changes in forest bird communities along aspen edges: The effect of aspen succession (abstract). *Abstracts, Fifth Annual Meeting of the Society for Conservation Biology,* Madison, Wisconsin, p. 82.

Matlack, G. 1994. Plant demography, land-use history, and the commercial use of forests. *Conservation Biology* 8(1): 298–299.

Matthiae, P.E. and F. Stearns. 1981. Mammals in forest lands in southeastern Wisconsin. Pages 55–66 in R.L. Burgess and D.M. Sharpe (eds.), *Forest Island Dynamics in Man-Dominated Landscapes*. New York: Springer-Verlag.

May, R.M. 1975. Patterns of species abundance and diversity. Pages 81–120 in M.L. Cody and J.M. Diamond (eds.), *Ecology and Evolution of Communities*. Cambridge, Massachusetts: Harvard University Press.

May, R.M. 1981. Models for single populations. Pages 5–29 in R.M. May (ed.), *Theoretical Ecology: Principles and Applications*. Boston: Blackwell Scientific Publications.

May, R.M. 1986. How many species are there? *Nature* 324: 514–515.

May, R.M. 1988. How many species are there on Earth? *Science* 241: 1441–1449.

McCaffery, K.R. 1986. On deer carrying capacity in northern Wisconsin. Pages 54–69 in R.J. Regan and S.R. Darling (compilers), *Transactions of the Twenty-second Northeast Deer Technical Committee*. Waterbury, Vermont: Vermont Fish and Wildlife Department.

McCoy, E.D. 1982. The application of island biogeographic theory to forest tracts: Problems in the determination of turnover rates. *Biological Conservation* 22: 217–227.

McKnight, W.N., Ed. 1993. *Biological Pollution: The Control and Impact of Invasive Exotic Species*. Indianapolis, Indiana: Indiana Academy of Science.

McShea, W.J. and J.H. Rappole. 1992. White-tailed deer as keystone species within forested habitats of Virginia. *Virginia Journal of Science* 43: 177–186.

Mech, L.D. 1966. *The Wolves of Isle Royale*. Washington, D.C.: U.S. Government Printing Office.

Mech, L.D., S.H. Fritts, G.L. Radde and W.J. Paul. 1988. Wolf distribution and road density in Minnesota. *Wildlife Society Bulletin* 16: 85–87.

Meine, C. 1988. *Aldo Leopold: His Life and Work*. Madison: The University of Wisconsin Press.

Menges, E. 1991. Seed germination percentage increases with population size in a fragmented prairie species. *Conservation Biology* 5: 158–164.

Menges, E.S. and D.M. Waller. 1983. Plant strategies in relation to elevation and light in floodplain herbs. *American Naturalist* 122: 454–473.

Merchant, C. 1989. *Ecological Restorations: Nature, Gender, and Science in New England*. Chapel Hill, North Carolina: University of North Carolina Press.

Middletown, J. and G. Merriam. 1985. The rationale for conservation: Problems from a virgin forest. *Biological Conservation* 33: 133–145.

Miller, S.G., S.P. Bratton and J. Hadidian. 1992. Impacts of white-tailed deer on endangered and threatened vascular plants. *Natural Areas Journal* 12(2): 67–74.

Mills, L.S. 1992. Edge effects on small mammals in forest remnants of southwest Oregon (abstract). *Northwest Science* 66(2): 127.

Mitchell, J.G. 1992. Love & war in the Big Woods. *Wilderness* 55 (spring): 11–22.

Mlot, C. 1992. Botanists sue Forest Service to preserve biodiversity. *Science* 257: 1618–1619.

Mohlenbrock, R.H. 1986. Tionesta Forest, Pennsylvania. *Natural History* 95(11): 74–78.

Moldenke, A. 1990. One hundred twenty thousand little legs. *Wings* (Summer): 11–14.

Moldenke, A. and J.D. Lattin. 1990a. Density and diversity of soil arthropods as biological probes of complex soil phenomena. *Northwest Environmental Journal* 6(2): 409–410.

Moldenke, A. and J.D. Lattin. 1990b. Dispersal characteristics of old-growth soil arthropods: The potential for loss of diversity and biological function. *Northwest Environmental Journal* 6(2): 408–409.

Mooney, H.A. and J.A. Drake, Eds. 1986. *Ecology of Biological Invasions of North America and Hawaii*. New York: Springer-Verlag.

Morey, H. F. 1936. A comparison of two virgin forests in northwestern Pennsylvania. *Ecology* 17(1): 43–55.

Morrison, P.H. and F.J. Swanson 1990. Fire history and pattern in a Cascade Mountain Landscape. *USDA Forest Service General Technical Report PNW-GRT-254*. Portland, Oregon: Pacific Northwest Research Station.

Muir, P.S. and J.E. Lotan. 1985. Disturbance history and serotiny of *Pinus contorta* in western Montana. *Ecology* 66: 1658–1668.

Murphy, D.D. and B.R. Noon. 1992. Integrating scientific methods with habitat conservation planning: Reserve design for northern spotted owls. *Ecological Applications* 20: 3–17.

Murphy, R.K., J.R. Cary, R.K. Anderson and N.F. Payne. 1986. Seasonal movements of white-tailed deer on declining habitats in central Wisconsin. *Transactions of the Wisconsin Academy of Sciences, Arts and Letters* 74: 133–146.

Nash, R.F. 1967. *Wilderness and the American Mind*. New Haven: Yale University Press (3rd ed. 1982).

Nash, T.H. and V. Wirth, Eds. 1988. *Lichens, Bryophytes, and Air Quality*. Bibliotheca Lichenologica. Berlin: J. Cramer.

National Academy of Sciences. 1979. *Animals as Monitors of Environmental Pollutants*. Washington, D.C.: National Academy Press.

National Academy of Sciences. 1992. *Report of a Workshop for a National Park Service Ecological Research Program*. Albuquerque, New Mexico: U.S. Department of the Interior.

Nee, S. and R.M. May. 1992. Dynamics of metapopulations: Habitat destruction and competitive coexistence. *Journal of Animal Ecology* 61(1): 37–40.

Newmark, W.D. 1987. A land-bridge island perspective on mammalian extinctions in western North American parks. *Nature* 325: 430–432.

Nicholls, A.O. and C.R. Margules. 1991. The design of studies to demonstrate the biological importance of corridors. Pages 49–61 in D.A. Saunders and R.J. Hobbs (eds.), *Nature Conservation 2: The Role of Corridors*. Chipping Norton, Australia: Surrey Beatty and Sons Pty Ltd.

Nichols, D. 1986. A new class of echinoderms. *Nature* 321: 808.

Nicholson, R. 1992. Death and *Taxus. Natural History* 101: 20-23.

Niemelä, J., D. Langor and J.R. Spence. 1993. Effects of clearcut harvesting on boreal ground-beetle assemblages (Coleoptera: Carabidae) in western Canada. *Conservation Biology* 7(3): 551–561.

Niering, W.A. 1992. The New England forests. *Restoration and Management Notes* 10(1): 24–28.

Nigh, T.A., W.L. Pflieger, P.L. Redfearn, Jr., W.A. Schroeder, A.R. Templeton and F.R. Thompson III. 1992. *The Biodiversity of Missouri: Definition, Status, and Recommendations for Its Conservation.* Jefferson City, Missouri: Missouri Department of Conservation.

Nilsson, I.N. and S.G. Nilsson 1982. Turnover of vascular plant species on small islands in Lake Mockeln, South Sweden, 1976–1980. *Oecologia* 53: 128–133.

Norse, E.A. 1989. *Ancient Forests of the Pacific Northwest: Sustaining Biological Diversity and Timber Production in a Changing World.* Washington, D.C.: Island Press.

Noss, R.F. 1987a. Do we really want diversity? *Whole Earth Review* (summer): 126–128.

Noss, R.F. 1987b. Protecting natural areas in fragmented landscapes. *Natural Areas Journal* 7(1): 2–13.

Noss, R.F. 1990. Indicators for monitoring biodiversity: A hierarchical approach. *Conservation Biology* 4(4): 355–364.

Noss, R.F. 1991a. Sustainability and wilderness. *Conservation Biology* 5(1): 120–122.

Noss, R.F. 1991b. Effects of edge and internal patchiness on avian habitat use in an old-growth Florida hammock. *Natural Areas Journal* 11(1): 34–47.

Noss, R.F. 1991c. Landscape connectivity: Different functions at different scales. Pages 27–39 in W.E. Hudson (ed.), *Landscape Linkages and Biodiversity.* Washington, D.C.: Island Press.

Noss, R.F. 1991d. From endangered species to biodiversity. Pages 227–246 in K.A. Kohm (ed.), *Balancing on the Brink of Extinction.* Washington, D.C.: Island Press.

Noss, R.F. 1992. The Wildlands Project—Land conservation strategy. Pages 10–25 in *Wild Earth.* Special Issue: "The Wildlands Project: Plotting a North American Wilderness Recovery Strategy." Canton, New York: Cenozoic Society, Inc.

Noss, R.F. 1993. A conservation plan for the Oregon Coast Range: Some preliminary suggestions. *Natural Areas Journal* 13(4): 276–290.

Noss, R.F. and A. Cooperrider. 1994. *Saving Nature's Legacy: Protecting and Restoring Biodiversity.* Washington, D.C.: Defenders of Wildlife and Island Press.

Nowacki, G.J. and P.A. Trianosky. 1993. Literature on old-growth forests of eastern North America. *Natural Areas Journal* 13(2): 87–107.

Oberwinkler, F. 1992. Biodiversity amongst filamentous fungi. *Biodiversity and Conservation* 1(4): 293–311.

O'Brien, M.H. 1990. NEPA as it was meant to be: *NCAP v. Block,* herbicides, and Region 6 Forest Service. *Environmental Law* 20: 735-745.

Oelschlaeger, M. 1991. *The Idea of Wilderness: From Prehistory to the Age of Ecology.* New Haven, Connecticut: Yale University Press.

Office of Technology Assessment. 1992. *Forest Service Planning: Accommodating Uses, Producing Outputs, and Sustaining Ecosystems.* Washington D.C.: U.S. Congress.

O'Neill, R.V., D.L. DeAngelis, J.B. Waide and T.F.H. Allen. 1986. *A Hierarchical Concept of Ecosystems*. Princeton, New Jersey: Princeton University Press.

Otis, D.L., K.P. Burnham, G. C. White and D.R. Anderson. 1978. Statistical inference from capture data on closed animal populations. *Wildlife Monographs* 62: 1–135.

O'Toole, R. 1988. *Reforming the Forest Service*. Washington, D.C.: Island Press.

Parker, L.R. 1990. *Feasibility Assessment for the Reintroduction of North American Elk, Moose, and Caribou into Wisconsin*. Madison: Department of Natural Resources.

Pedrós-Alió, C. 1993. Diversity of bacterioplankton. *Trends in Ecology and Evolution* 8(3): 86–90.

Peek, J.M. 1986. *A Review of Wildlife Management*. Englewood Cliffs, New Jersey: Prentice Hall.

Pelton, M. 1985. Habitat needs of black bears in the East. Pages 49–53 in D.L. Kulhavy and R.M. Conner (eds.), *Wilderness and Natural Areas in the Eastern United States: A Management Challenge*. Nacogdoches, Texas: Center for Applied Studies, School of Forestry, Austin State University.

Perlin, J. 1989. *A Forest Journey: The Role of Wood in the Development of Civilization*. New York: W.W. Norton.

Perry, D.A., M.P. Amaranthus, J.G. Borchers, L.L. Borchers and R.E. Brainerd. 1989. Bootstrapping in ecosystems. *BioScience* 39: 230–236.

Peters, R. and J. Darling. 1985. The greenhouse effect and the nature reserves. *BioScience* 35: 707–717.

Peterson, C.J. and S.T. A. Pickett. 1991. Treefall and resprouting following catastrophic windthrow in an old-growth hemlock-hardwoods forest. *Forest Ecology and Management* 42: 205–217.

Phillips, D.J.H. 1980. *Quantitative Aquatic Biological Indicators*. London: Applied Science Publishers.

Phillips, D.L. and D.J. Shure. 1990. Patch-size effects on early succession in southern Appalachian forests. *Ecology* 71: 204–212.

Phillips, O. 1993. The potential for harvesting fruits from tropical rainforests: New data from Amazonian Peru. *Biodiversity and Conservation* 2: 18–39.

Pianka, E.R. 1983. *Evolutionary Ecology*. New York: Harper and Row.

Pickett, S. and J. Thompson. 1978. Patch dynamics and the design of nature reserves. *Biological Conservation* 13: 27–37.

Pils, C.M. 1983. Fisher (*Martes pennanti*) *Wisconsin Department of Natural Resources, Furbearer Profiles Report #3*. Madison: Department of Natural Resources.

Pinchot, G. 1947. *Breaking New Ground*. New York: Harcourt Brace and Co. (Reprint: Washington, D.C.: Island Press, 1987.)

Pinkett, H.T. 1970. *Gifford Pinchot, Public and Private Forester*. Urbana, Illinois: University of Illinois Press.

Porter, W.F. 1992. High-fidelity deer. *Natural History* 5/92: 48–49.

Post, S.L. 1991. Native Illinois species and related bibliography, Appendix 1. Pages 463–475 in *Illinois Natural History Survey Bulletin 34*, Article 4.

Postel, S. and J.C. Ryan. 1991. Reforming forestry. Pages 74–92 in L. Starke (ed.), *State of the World 1991*. New York: W.W. Norton.

Powell, R.A. 1982. *The Fisher: Life History, Ecology and Behavior*. Minneapolis, Minnesota: University of Minnesota Press.

Pöyry, J. 1992. Forest soils: A technical paper for a generic environmental impact statement on timber harvesting and forest management in Minnesota. St. Paul, Minnesota: Minnesota Environmental Quality Board.

Preston, F.W. 1962a. The canonical distribution of commonness and rarity: Part I. *Ecology* 43: 185–215.

Preston, F.W. 1962b. The canonical distribution of commonness and rarity: Part II. *Ecology* 43: 410–432.

Preston, R. 1992. Crisis in the hot zone. *New Yorker* 68(Jan.): 58-81.

Pyne, S.J. 1982. *Fire in America: A Cultural History of Wild Land and Rural Fire*. Princeton, New Jersey: Princeton University Press.

Pyne, S.J. 1989. The summer we let wild fire loose. *Natural History* (Aug.): 45–49.

Ralls, K. and J. Ballou. 1982. Effects of inbreeding on infant mortality in captive primates. *International Journal of Primatology* 3: 491–505.

Ralls, K., K. Brugger and J. Ballou. 1979. Inbreeding and juvenile mortality in small populations of ungulates. *Science* 206: 1101–1103.

Ranney, J.W., M.C. Bruner and J.B. Levenson. 1981. The importance of edge in the structure and dynamics of forest islands. Pages 67–96 in R.L. Burgess and D.M. Sharpe (eds.), *Forest Island Dynamics in Man-Dominated Landscapes*. New York: Springer-Verlag.

Raven, P.H. and G.B Johnson. 1986. *Biology*. St. Louis: Times Mirror/Mosby College.

Raven, P.H. and E.O. Wilson. 1992. A fifty-year plan for biodiversity surveys. *Science* 258: 1099–1100.

Read, D.J., D.H. Lewis, A.H. Fitter and I.J. Alexander. 1992. *Mycorrhizas in Ecosystems*. Wallingford, U.K.: C-A-B International.

Reese, K.P. and J.T. Ratti. 1988. Edge effect: A concept under scrutiny. Pages 127–136 in R.E. McCabe (ed.), *Transactions of the Fifty-third North American Wildlife Natural Resources Conference*. Washington, D.C.: Wildlife Management Institute.

Reibesell, J.F. 1982. Arctic-alpine plants on mountaintops: Agreement with island biogeography theory. *American Naturalist* 119: 657–674.

Rey, J.R. 1981. Ecological biogeography of arthropods on *Spartina* islands in northwest Florida. *Ecological Monographs* 51: 237–265.

Ricker, W.E. 1975. Computation and interpretation of biological statistics of fish populations. *Fisheries Research Board of Canada, Bulletin 191*.

Ricklefs, R.E. 1993. Affinities and rationales. *Science* 259: 1774–1775.

Risser, P.G. and J. Lubchenco, Eds. 1992. *Report of a Workshop for a National Park Service Ecological Research Program*. Albuquerque, New Mexico: U.S. Department of Interior.

Robbins, C.S., D.K. Dawson and B.A. Dowell. 1989. *Habitat Area Requirements of Breeding Forests Birds of the Middle Atlantic States*. Wildlife Monographs 103. Bethesda, Maryland: The Wildlife Society.

Roberts, R. 1993. In my view. *Paper News* (first quarter): 18–19 (Georgia Pacific Corporation, Port Edwards, Wisconsin).

Robertson, F.D. 1992. USDA to eliminate clearcutting as standard practice on National Forests. *USDA News Release*, June 4, 1992. Washington, D.C.: Forest Service.

Robinson, G. 1988. *The Forest and the Trees: A Guide to Excellent Forestry.* Washington, D.C.: Island Press.

Robinson, G.O. 1975. *The Forest Service: A Study in Public Land Management.* Baltimore, Maryland: Johns Hopkins University Press.

Robinson, S.K. 1988. Reappraisal of the costs and benefits of habitat heterogeneity for non-game wildlife. Pages 145–155 in R.E. McCabe (ed.), *Transactions of the Fifty-third North American Wildlife and Natural Resources Conference.* Washington, D.C.: Wildlife Management Institute.

Robinson, S.K. 1992. Population dynamics of breeding neotropical migrants in a fragmented Illinois landscape. Pages 408–418 in J.M. Hagan and D.W. Johnston (eds.), *Ecology and Conservation of Neotropical Migrant Landbirds.* Washington, D.C.: Smithsonian Institution Press.

Rodgers, A.D. 1968. *Bernhard Eduard Fernow: A Story of North American Forestry.* New York: Hafner Publishing Company.

Rogers, L.L. 1987. *Effects of Food Supply and Kinship on Social Behavior, Movements and Population Growth of Black Bears in Northeastern Minnesota.* Wildlife Monographs 97. Bethesda, Maryland: The Wildlife Society.

Roland, J. 1993. Large-scale forest fragmentation increases the duration of tent caterpillar outbreak. *Oecologia* 93: 25–30.

Rudnicky, T.C. and M.L. Hunter, Jr. 1993. Reversing the fragmentation perspective: Effects of clearcut size on bird species richness in Maine. *Ecological Applications* 3(2): 357–366.

Runkle, J.R. 1982. Patterns of disturbance in some old-growth mesic forests of eastern North America. *Ecology* 63(5): 1533–1546.

Runkle, J.R. 1985. Disturbance regimes in temperate forests. Page 472 in S.T.A. Pickett and P.S. White (eds.), *The Ecology of Natural Disturbance and Patch Dynamics.* Orlando, Florida: Academic Press.

Runkle, J.R. 1991. Gap dynamics of old-growth eastern forests: Management implications. *Natural Areas Journal* 11: 19–25.

Rusterholz, K.A. 1989. Old-growth forests in Minnesota: A preliminary report. *Biological Report No. 5, Minnesota Department of Natural Resources.*

Salwasser, H. 1987. Editorial. *Conservation Biology* 1: 275–277.

Salwasser, H. 1991a. Perspectives on sustainable forestry and the National Forest System. Text of speech, 11 April 1991, Spring Lecture Series, Issues in Agricultural and Life Sciences, School of Natural Resources, University of Wisconsin, Madison.

Salwasser, H. 1991b. New Perspectives for sustaining diversity in the U.S. National Forest ecosystems. *Conservation Biology* 5(4): 567–569.

Salwasser, H. 1992. From New Perspectives to Ecosystem Management: Response to Frissell et al. and Lawrence and Murphy. *Conservation Biology* 6: 469-472.

Sample, V.A. 1990. *The Impact of the Federal Budget Process on National Forest Planning.* New York: Greenwood Press.

Sample, V.A. 1991. *Land Stewardship in the Next Era of Conservation.* Milford, Pennsylvania: Grey Towers Press.

Sargent, C.S. 1933. *Manual of the Trees of North America.* New York: Houghton-Mifflin Company.

Saunders, D.A. and R.J. Hobbs, Eds. 1991. *Nature Conservation 2: The Role of Corridors.* Chipping Norton, Australia: Surrey Beatty & Sons Pty Ltd.

Saunders, D.A., R.J. Hobbs and C.R. Margules. 1991. Biological consequences of ecosystem fragmentation: A review. *Conservation Biology* 5: 18–32.

Schoener, T.W. and A. Schoener. 1983. The time to extinction of a colonizing propagule of lizards increases with island area. *Nature* 302: 332–334.

Schorger, A.W. 1965. The beaver in early Wisconsin. *Transactions of the Wisconsin Academy of Sciences* 54: 147–179.

Schowalter, T.D. 1986. Ecological strategies of forest insects and the need for a community-level approach to reforestation. *New Forestry* 1: 57–66.

Schowalter, T.D. 1989. Canopy arthropod community structure and herbivory in old-growth and regenerating forests in western Oregon. *Canadian Journal of Forestry Research* 19: 318–322.

Schowalter, T.D. and D.A. Crossley, Jr. 1987. Canopy arthropods and their response to forest disturbance. Pages 207–218 in W.T. Swank and D.A. Crossley (eds.), *Forest Hydrology and Ecology at Coweeta.* New York: Springer-Verlag.

Schowalter, T.D., J.W. Webb and D.A. Crossley, Jr. 1981. Community structure and nutrient content of canopy arthropods in clearcut and uncut forest ecosystems. *Ecology* 62: 1010–1019.

Scott, J.M., B. Csuti and S. Caicco. 1991. Gap analysis: Assessing protection needs. Pages 15–26 in W.E. Hudson (ed.), *Landscape Linkages and Biodiversity.* Washington, D.C.: Island Press.

Shafer, C.L. 1990. *Nature Reserves: Island Theory and Conservation Practice.* Washington, D.C.: Smithsonian Institution Press.

Shaw, C.H., H. Lundkvist, A. Moldenke and J.R. Boyle. 1990. The relationships of soil fauna to long-term forest productivity in temperate and boreal ecosystems: Processes and research strategies. Pages 39–77 in W.J. Dyck and C.A. Mees (eds.), *Long-Term Field Trials to Assess Environmental Impacts of Harvesting.* IEA/BE T6/A6 Report No. 5, Forest Research Institute. Rotorua, New Zealand: FRI Bulletin No. 161.

Sheldon, H.L. 1984. Procedure for determining and maintaining viable populations of wildlife species on the Chequamegon National Forest. *Forest Planning Documentation Record.* Chequamegon National Forest, Park Falls, Wisconsin.

Shepherd, J. 1975. *The Forest Killers: The Destruction of the American Wilderness.* New York: Weybright and Talley.

Shugart, H.H. and D.C. West. 1977. Development of an Appalachian deciduous forest succession model and its application to assessment of the impact of the chestnut blight. *Journal of Environmental Management* 5: 161–179.

Shugart, H.H. and D.C. West. 1980. Forest succession models. *BioScience* 30: 308–313.

Siegel, W.C. 1990. *Legislative regulation of private forestry practices in the United States—Recent trends*. Forestry Legislation, Zurich, Switzerland: International Union of Forestry Research Organizations.

Simberloff, D.S. 1987. The spotted owl fracas: Mixing academic, applied, and political ecology. *Ecology* 68: 766–772.

Simberloff, D.S. and L.G. Abele. 1982. Refuge design and island biogeographic theory: Effects of fragmentation. *American Naturalist* 120: 41–50.

Simberloff, D.S., J.A. Farr, J. Cox and D.W. Mehlman. 1992. Movement corridors: Conservation bargains or poor investments? *Conservation Biology* 6(4): 493–504.

Simpson, G.G. 1984. Mammals and cryptozoology. *Proceedings of the American Philosophical Society* 128(1): 1–19.

Skole, D. and C. Tucker. 1993. Tropical deforestation and habitat fragmentation in the Amazon: Satellite data from 1978 to 1988. *Science* 260: 1905–1910.

Skow, J. 1988. The forest service follies. *Sports Illustrated* 68: 74-88.

Small, M.F. and M.L. Hunter, Jr. 1989. Response of passerines to abrupt forest-river and forest-powerline edges in Maine. *Wilson Bulletin* 101(1): 77–83.

Smith, S. 1990. Afterlife of a whale. *Discover* 2/90: 46–49.

Smith, T.L. 1989. An overview of old-growth forests in Pennsylvania. *Natural Areas Journal* 9(1): 40–44.

Solheim, S.L., D.M. Waller and W.S. Alverson. 1987. Maintaining biotic diversity in national forests: The necessity for large blocks of mature forest. *Endangered Species Technical Bulletin Reprint* 4(8): 1–3.

Solheim, S.L., D.M. Waller and W.S. Alverson. 1991. *Inventory and monitoring in the Chequamegon National Forest*. Unpublished report prepared for the Office of Technology Assessment of the U.S. Congress, 31 pp. Dept. of Botany, University of Wisconsin, Madison.

Sollins, P. and F.M. McCorison. 1981. Nitrogen and carbon solution chemistry of an old growth coniferous forest watershed before and after cutting. *Water Resources Research* 17: 1409–1418.

Soulé, M. and D. Simberloff. 1986. What do genetics and ecology tell us about the design of nature preserves? *Biological Conservation* 35: 19–40.

Spellerberb, I.F. 1992. *Evaluation and Assessment for Conservation*. London: Chapman and Hall.

Sprugel, D.G. 1976. Dynamic structure of wave-generated *Abies balsamea* forests in the northeastern United States. *Journal of Ecology* 64: 889–911.

Spurr, S.H. 1956. Natural restocking of forests following the 1938 hurricane in central New England. *Ecology* 30: 350–358.

Stanosz, G.R. and R.F. Patton. 1987. Armillaria root rot in aspen stands after repeated short rotations. *Canadian Journal of Forestry Research* 17: 1001–1005.

St. Clair, J. 1991. Crackdown in North Carolina. *Forest Watch* March 1991: 10-11.

Steen, H.K. 1976. *The U.S. Forest Service: A History*. Seattle, Washington: University of Washington Press.

Steen, H.K., Ed. 1984. *History of Sustained-Yield Forestry.* Durham, North Carolina: Forest History Society.

Stevens, W.K. 1993a. Babbit to map ecosystems under policy shift. *The New York Times,* March 14, 1993, p. A14.

Stevens, W.K. 1993b. Biologists fear sustainable yield is unsustainable idea. *The New York Times,* April 20, 1993, p. B10.

Stoddart, L.A., A.R. Smith and T.W. Box. 1975. *Range Management.* New York: McGraw-Hill.

Stutz, B. 1993. Stands of time. *Audubon* 95(1): 62–77.

Sutherland, F.R. and R. Beers. 1991. Supreme indifference. *The Amicus Journal* 13: 38-42.

Swain, A.M. 1973. A history of fire and vegetation in northeastern Minnesota as recorded in lake sediments. *Quaternary Research* 3: 383–396.

Temple, S.A. 1986. Ecological principles of wildlife management. Pages 11–21 in J.B. Hale, L.B. Best and R.L. Clawson (eds.), *Management of Nongame Wildlife in the Midwest: A Developing Art.* Proceedings of the Forty-seventh Midwest Fish and Wildlife Conference, Grand Rapids, Michigan.

Temple, S.A. 1988. When is a bird's habitat not habitat? *Passenger Pigeon* 50: 37–42.

Temple, S.A. 1990. The nasty necessity: Eradicating exotics. *Conservation Biology* 4: 113–115.

Templeton, A. and B. Read. 1984. Factors eliminating inbreeding depression in a captive herd of Speke's Gazelle (*Gazella spekei*). *Zoo Biology* 3: 177–199.

Terborgh, J. 1974. The preservation of natural diversity: The problem of extinction prone species. *BioScience* 24: 715–722.

Terborgh, J. 1989. *Where Have All the Birds Gone?* Princeton, New Jersey: Princeton University Press.

Terborgh, J. 1992. Why American songbirds are vanishing. *Scientific American* 266: 56–62.

Thiel, R.P. 1985. Relationship between road densities and wolf habitat suitability in Wisconsin. *American Midland Naturalist* 11: 404–407.

Thomas, J.W. and H. Salwasser. 1989. Bringing conservation biology into a position of influence in natural resource management. *Conservation Biology* 3(2): 123–127.

Thomas, J.W., C. Maser and J.E. Rodiek. 1979. Edges. Pages 48–59 in J.W. Thomas (ed.), *Wildlife Habitats in Managed Forest: The Blue Mountains of Oregon and Washington.* USDA Forest Service, Agriculture Handbook No. 553, Washington, D.C.

Thompson, F.R., III, W.D. Dijak, T.G. Kulowiec and D.A. Hamilton. 1992. Breeding bird populations in Missouri Ozark forests with and without clearcutting. *Journal of Wildlife Managment* 56(1): 23–30.

Thomson, J.W. 1990. Lichens in old-growth woods in Wisconsin. *Bulletin of the Wisconsin Botanical Club* 22(1): 7–10.

Thornton, P.L. A balanced program for the National Forest System. 1974. Pages 360–366 in J. B. Trefethen and K. J. Sabol (eds.), *Transactions of the Thirty-*

ninth North American Wildlife and Natural Resources Conference. Washington, D.C.: Wildlife Management Institute.

Tierson, W.C., G.F. Mattfeld, R.W. Sage, Jr. and D.F. Behrend. 1985. Seasonal movements and home ranges of white-tailed deer in the Adirondacks. *Journal of Wildlife Managment* 49(3): 760–769.

Tilghman, N.G. 1989. Impacts of white-tailed deer on forest regeneration in northwestern Pennsylvania. *Journal of Wildlife Managment* 53(3): 524–532.

Tilghman, N.G. and K.E. Evans. 1986. A framework for nongame management in midwestern states. Pages 97–115 in J.B. Hale, L.B. Best and R.L. Clawson (eds.), *Management of Nongame Wildlife in the Midwest: A Developing Art.* Proceedings of the Forty-seventh Midwest Fish and Wildlife Conference, Grand Rapids, Michigan.

Trippensee, E. 1948. *Wildlife Management.* New York: McGraw-Hill.

Trüper, H.G. 1992. Prokaryotes: An overview with respect to biodiversity and environmental importance. *Biodiversity and Conservation* 1(4): 227–236.

Turner, M.G., Ed. 1987. *Landscape Heterogeneity and Disturbance.* New York: Springer-Verlag.

Turner, M.G. and V.H. Dale. 1991. Modeling landscape disturbance. Pages 323–351 in M.G. Turner and R.H. Gardner (eds.), *Quantitative Methods in Landscape Ecology.* New York: Springer-Verlag.

Turner, M.G., W.H. Romme, R.H. Gardner, R.V. O'Neill and T.K. Kratz. 1993. A revised concept of landscape equilibrium: Disturbance and stability on scaled landscapes. *Landscape Ecology* 8: 213–227.

Turner, T. 1990. *Wild by Law: The Sierra Club Legal Defense Fund and the Places It Has Saved.* San Francisco: Sierra Club Books.

Tyrrell, L.E. 1992. Characteristics, distribution and management of old-growth forests on units of the U.S. National Park Service: Results of a questionnaire. *Natural Areas Journal* 12(4): 198–205.

Udall, S.L. 1963. *The Quiet Crisis.* New York: Holt, Rinehart and Winston.

USDA Forest Service. 1986a. *Chequamegon National Forest Land and Resource Management Plan, Final Environmental Impact Statement and Record of Decision.* Park Falls, Wisconsin: USDA Forest Service, Chequamegon National Forest.

USDA Forest Service. 1986b. *Nicolet National Forest Land and Resource Management Plan, Final Environmental Impact Statement and Record of Decision.* Rhinelander, Wisconsin: USDA Forest Service, Nicolet National Forest.

USDA Forest Service. 1986c. *Allegheny National Forest Land and Resource Management Plan, Final Environmental Impact Statement and Record of Decision.* Warren, Pennsylvania: USDA Forest Service, Allegheny National Forest.

USDA Forest Service. 1991. Advance notice of proposed rulemaking. *56 Federal Register 6508 et seq.* February 15, 1991. Washington, D.C.: U.S. Government Printing Office.

USDA Forest Service. 1993. Lichens as bioindicators of air quality. *General Technical Report RM-224.* Washington, D.C.: USDA Forest Service.

USDA Forest Service, Eastern Region. 1992a. Ecosystems management—New Perspectives strategic plan, Milwaukee, Wisconsin: USDA Forest Service. (Document dated January 16, 1992).

USDA Forest Service, Eastern Region. 1992b. Ecosystem management strategies for the northeastern and midwestern national forests: A report to the Chief of the Forest Service. Rhinelander, Wisconsin: USDA Forest Service North Central Experiment Station. (Document dated September 4, 1992).

USDA Forest Service and USDI Bureau of Land Management. 1993. *Draft Supplemental Environmental Impact Statement on Management of Habitat for Late-successional and Old-growth Forest Related Species Within the Range of the Northern Spotted Owl.* Publication 1993—793-234/82403. Washington D.C.: U.S. Government Printing Office.

USFWS (U.S. Fish and Wildlife Service, Department of Interior). 1978. *Eastern Timber Wolf Recovery Plan.* Washington, D.C.: U.S. Government Printing Office.

Van Dung, V., P. Mong Giao, N. Ngoc Chinh, D. Tuoc, P. Arctander and J. MacKinnon. 1993. A new species of living bovid from Vietnam. *Nature* 363: 443–445.

Van Dyke, J.M., D. Zaelke and G. Hewison. 1993. *Freedom for the Seas in the 21st Century.* Washington, D.C.: Island Press.

Vasquez, R. and A. Gentry. 1989. Use and misuse of forest-harvested fruits in the Iquitos area. *Conservation Biology* 3(4): 350–361.

Verme, L.J. 1973. Movements of white-tailed deer in upper Michigan. *Journal of Wildlife Management* 37(4): 545–552.

Vickerman, K. 1992. The diversity and ecological significance of protozoa. *Biodiversity and Conservation* 1(4): 334–341.

Waller, D.M. 1991. Introduction. Pages 3–13 in W.E. Hudson (ed.), *Landscape Linkages and Biodiversity.* Washington, D.C.: Island Press.

Waller, D.M. 1993a. The statics and dynamics of mating system evolution. In N. Thornhill (ed.), *The Natural History of Inbreeding and Outbreeding.* Chicago: Univeristy of Chicago Press.

Waller, D.M. 1993b. Wisconsin's Scientific Roundtable: Uniting research and management. *Inner Voice* 5(3): 13.

Walters, B.B. 1991. Small mammals in a subalpine old-growth forest and clearcuts. *Northwest Science* 65(1): 27–31.

Walters, C.J. 1986. *Adaptive Management of Renewable Resources.* New York: MacMillan.

Wayne, R.K., N. Lehman, D. Girman, P.J.P. Gogan, D.A. Gilbert, K. Hansen, R.O. Peterson, U.S. Seal, A. Eisenhawer, L.D. Mech and R.J. Krumenaker. 1991. Conservation genetics of the endangered Isle Royale gray wolf. *Conservation Biology* 5: 41–51.

Westman, W.E. 1990. Managing for biodiversity. *BioScience* 40: 26–33.

WFCTF (Wisconsin Forest Conservation Task Force). 1986. Statement of reasons in support of appeals of the long-range management plan for the Chequamegon and Nicolet National Forests. Madison, Wisconsin: Wisconsin Audubon and John Muir Chapter of the Sierra Club.

Whetmore, C. 1993. Lichens and air quality in the Round Lake Wilderness. Unpublished report submitted to the Chequamegon National Forest, Park Falls, Wisconsin.

Whitcomb, R.F., C.S. Robbins, J.F. Lynch, B.L. Whitcomb, M.K. Klimkiewicz and D. Bystrak. 1981. Effects of forest fragmentation on avifauna of the eastern deciduous forest. Pages 125–205 in R.L. Burgess and D.L. Sharpe (eds.), *Forest Island Dynamics in Man-Dominated Landscapes*. New York: Springer-Verlag.

White, P.S. and S.T.A. Pickett, Eds. 1985. Natural disturbance and patch dynamics: An introduction. In *The Ecology of Natural Disturbance and Patch Dynamics*. New York: Academic Press.

White, P.S., R.I. Miller and S.P. Bratton. 1983. Island biogeography and preserve design: Preserving the vascular plants of Great Smoky Mountain National Park. *Natural Areas Journal* 3: 4–13.

Whitney, G.G. 1984. Fifty years of change in the arboreal vegetation of Heart's Content, an old-growth hemlock–white pine–northern hardwood stand. *Ecology* 65: 403–408.

Whittaker, R.H. 1972. Evolution and measurement of species diversity. *Taxon* 21: 213–251.

Whittaker, R.H. 1975. *Communities and Ecosystems*. New York: Macmillian.

Wilcove, D.S. 1985. Nest predation in forest tracts and the decline of migratory songbirds. *Ecology* 66: 1211–1214.

Wilcove, D.S. 1988a. Forest fragmentation as a wildlife management issue in the eastern United States. Pages 1–5 in R.M. DeGraaf and W.M. Healy (eds.), *Is Forest Fragmentation a Management Issue in the Northeast?* Rochester, New York: USDA Forest Service Northeastern Forest Experiment Station.

Wilcove, D.S. 1988b. *National Forests: Policies for the Future. Vol. 2: Protecting Biological Diversity*. Washington, D.C.: The Wilderness Society.

Wilcove, D.S. 1989. Protecting biodiversity in multiple-use lands: Lessons from the U.S. Forest Service. *Trends in Ecology and Evolution* 4: 385–388.

Wilcove, D.S., C.H. McLellan and A.P. Dobson. 1986. Habitat fragmentation in the temperate zone. Pages 237–256 in M.E. Soulé (ed.), *Conservation Biology: The Science of Scarcity and Diversity*. Sunderland, Massachusetts: Sinauer Associates.

Wilcox, B.A. 1978. Supersaturated island faunas: A species-age relationship for lizards on post-Pleistocene land-bridge islands. *Science* 19: 996–998.

Wilcox, B.A. 1980. Insular ecology and conservation. Pages 95–117 in M.E. Soulé and B.A. Wilcox (eds.), *Conservation Biology: An Evolutionary–Ecological Perspective*. Sunderland, Massachusetts: Sinauer Associates.

Wilcox, B.A., D.D. Murphy, P.R. Ehrlich and G.T. Austin. 1986. Insular biogeography of the montane butterfly faunas of the Great Basin: Comparison with birds and mammals. *Oecologia* 69: 188–194.

Wilkinson, C.F. and H.M. Anderson. 1985. Land and resource planning in the National Forests. *Oregon Law Review* 64: 1–373. Reprinted 1987, Island Press, under the same title.

Willers, B. 1992. Toward a science of letting things be. *Conservation Biology* 6(4): 605–607.

Williams, M. 1989. *Americans and Their Forests: A Historical Geography.* New York: Cambridge University Press.

Wilson, A.N. 1961. The nature of the taxon cycle in the Melanesian ant fauna. *American Naturalist* 95: 169–193.

Wilson, E.O. 1984. *Biophilia.* Cambridge, Massachusetts: Harvard University Press.

Wilson, E.O. 1988. *Biodiversity.* Washington, D.C.: National Academy Press.

Wilson, E.O. 1989. Threats to biodiversity. *Scientific American* 261(3): 108–116.

Wilson, E.O. 1992. *The Diversity of Life.* New York: W.W. Norton.

Wilson, J.B. and A.D.Q. Agnew. 1992. Positive feedback switches in plant communities. *Advances in Ecological Research* 23: 263–336.

Winckler, S. 1992. Stopgap measures. *The Atlantic Monthly* 269: 74–81.

World Resources Institute. 1992. *The 1992 Information Please Environmental Almanac.* Washington, D.C.: World Resources Institute.

Wren, C.D. 1986. Mammals as biological indicators of environmental metal levels. *Environmental Monitoring and Assessment* 6: 127–144.

Wright, S. 1931. Evolution in Mendelian populations. *Genetics* 10: 97–159.

Wright, S. 1982. Character change, speciation, and the higher taxa. *Evolution* 36: 427–443.

Yahner, R.H. 1988. Changes in wildlife communities near edges. *Conservation Biology* 2(4): 333–339.

Yahner, R.H. and D.P. Scott. 1988. Effects of forest fragmentation on depredation of artificial nests. *Journal of Wildlife Managment* 52: 158–161.

Yoon, C.K. 1993. Counting creatures great and small. *Science* 260: 620–622.

Young, D.K. 1988. The marsh beetles (Coleoptera: Scirtidae) of Pine Hollow and the UW-Milwaukee field station. *UW-Milwaukee Field Station Bulletin* 21: 1–7.

Zuckerman, S. 1992. New forestry or new hype? *Sierra* (Mar/Apr): 41–45, 67.

Glossary of Abbreviations and Acronyms

AFSEEE Association of Forest Service Employees for Environmental Ethics, an independent, vocal group dissatisfied with current leadership within the Forest Service; founded by Jeff DeBonis.

Agency When used alone and capitalized (not in combination with a specific name), we are referring to the USDA Forest Service.

ANPR Advance Notice of Proposed Rulemaking, an initial step in the federal regulation promulgation process under the Administrative Procedure Act. Here, referring to the proposal of the Forest Service to rewrite the NFMA planning regulation, 36 C.F.R. Part 219, published 2/15/91 at 56 Federal Register 6508.

ASQ Allowable Sale Quantity, the maximum amount of timber allowed to be sold in a year on any given National Forest. These controversial goals are usually determined by priorities and budgets set within National Forests and Forest Service Regions, a process not immune to pressure from the Washington office and political pressure.

BLM Bureau of Land Management, an agency within the Department of Interior.

DMA Diversity Maintenance Area, a large block of contiguous habitat formally designated for the primary purpose of maintaining biological diversity by limiting the extent and severity of anthropogenic disturbance. First proposed in 1985 for the Wisconsin National Forests (see Chapters 10 and 11).

DNR Department of Natural Resources, the state agency usually responsible for forestry, wildlife, and other land management decisions.

DUZ Dominant-Use Zoning Model of forest management (see Chapters 10 and 11).

ECS Ecological Classification System, a hierarchical system used by the Forest Service for categorizing forest communities according to their soil, landform, vegetative composition, and climate. The types extend from broad-scale "province" and "section" levels through Land Type Associations (LTA), Ecological Land Types (ELT), and Ecological Land Type Phase (ELTP) to Site. While the system is intended to be synthetic and inclusive, it originated from schemes used to type soils to predict timber growing capability (see Chapter 7).

EIS Environmental Impact Statement, a required public disclosure document under NEPA. A Final EIS (FEIS) accompanied the issuance of the Forest Plans for each National Forest. Also, Draft EIS (DEIS) and Draft Supplemental EIS (DSEIS).

EPA   Environmental Protection Act of 1970. Also, the Environmental Protection Agency, advanced to cabinet status under President Clinton in 1993.

ESA   Endangered Species Act of 1973 (16 U.S.C. §§ 1531 et seq.); up again for reauthorization in 1994.

GIS   Geographic Information System, a database accessed and processed geographically and usually containing multiple discrete "layers" composed of different types of data (e.g., land ownership, forest community type, roads, etc.). GIS approaches are gaining momentum for their ability to cope with intrinsically geographic information, such as the arrangement of timber harvests or the shape and proximity of reserved natural areas.

hectare (ha)   Metric unit of areal measure, equal to 2.471 acres.

kilometer (km)   Metric unit of distance, equal to 0.621 miles.

K-V   Knutson-Vandenberg Act of 1930, which allowed the USFS to retain a proportion of timber receipts for operations related to reforestation, such as site preparation. The NFMA expanded use of these funds to include wildlife management and other activities. K-V funds act as an incentive to favor timber harvests by amounting to a sizable proportion of total budgets within individual Forests (see O'Toole 1988).

MCA or MDA   Minimum Critical (or Dynamic) Area, that threshold area thought to be minimally sufficient to maintain natural disturbance regimes and predictable recolonization by populations of species dependent on metapopulation dynamics for persistence.

MIS   Management Indicator Species, monitoring of which is intended to reflect the response of a forest's biota to management.

MUM   The Multiple Use Module Model of forest management developed by Larry Harris and Reed Noss (see Chapter 10).

MUSY   Multiple-Use Sustained-Yield Act of 1960; requires each National Forest to cut an amount each year that is no greater than what can be sustained on a nondeclining basis indefinitely. Also explicitly recognizes the legitimacy of recreation, wildlife, grazing, and timber as legitimate uses.

NBS   National Biological Survey; division of the Department of the Interior established by Secretary of the Interior Bruce Babbitt in October 1993 to address in a more systematic fashion threats to biodiversity across the U.S. Initially made up primarily of scientific research staff from the National Park Service and the U.S. Fish and Wildlife Service.

NCFES   North Central Forest Experiment Station, with offices in St. Paul, MN, and Rhinelander, WI.

NEPA   National Environmental Policy Act of 1970; established the requirement to prepare environmental impact statements (EISs) for major federal activities and those requiring federal permits.

NFMA   National Forest Management Act of 1976.

NPS   National Park Service, within the Departmemt of Interior.

OA   Opportunity Area, a land unit for planning within National Forests larger than a stand and smaller than a district.

OMB   Office of Management and Budget, federal administration.

region   One of nine large-scale planning and administrative units within the
National Forest System; encompasses several states. Region 9, for example, constitutes the Eastern Region and covers the Northeast and upper Midwest. It is
headquartered in Milwaukee, WI.

RNA   Research Natural Area, an area set aside specifically to preserve significant
natural features, including rare habitats and associated rare and threatened plant
and animal species.

RPA   Forest and Rangeland Renewable Resources Planning Act of 1974 (16
U.S.C. § 1600 et seq.).

SAF   Society of American Foresters, the chief professional organization for
foresters in North America.

SLOSS   "Single Large or Several Small" debate over how best to design nature reserves to retain diversity.

SOR   Statement of Reasons, usually in reference to the documents drawn up by
the environmental appellants to appeal the long-range management plans of the
Wisconsin National Forests.

square kilometer (km²)   Metric unit of areal measure, equal to 0.386 square miles.

USDA   U.S. Department of Agriculture, home of the U.S. Forest Service.

USDI   U.S. Department of the Interior, home to the Park Service, the U.S. Fish
and Wildlife Service, and the Bureau of Land Management.

USFS or FS   U.S. Forest Service, within the Department of Agriculture.

WFCTF   Wisconsin Forest Conservation Task Force, a small group of environ-
mentalists and scientists that first came together in 1985 to petition the Forest
Service to revise its management plans for the Chequamegon and Nicolet
National Forests.

Species List

alewife—*Alosa pseudoharengus*

alfalfa—*Medicago sativa* L.

American burying beetle—*Nicrophorus americanus*

aspen, quaking aspen—*Populus tremuloides* Michx.

Atlantic salmon—*Salmo salar*

balsam fir—*Abies balsamea* (L.) Mill.

basswood—*Tilia americana* L.

bear, black bear—*Ursus americanus*

beaver—*Castor canadensis*

beech, American beech—*Fagus grandifolia* Ehrh.

beech bark disease—*Nectria* species complex

beech scale insects—*Cryptococcus fagisuga* Lindinger

birches—*Betula* species

bison, woodland bison—*Bison bison*

black-and-white warbler—*Mniotilta varia*

Blackburnian warbler—*Dendroica fusca*

black locust—*Robinia pseudoacacia* L.

black spruce—*Picea mariana* (Mill.) BSP

blue jay—*Cyanocitta cristata*

bobcat—*Felis rufus*

brainworm—*Parelaphostrongylus tenuis*

Braun's holly fern—*Polystichum braunii* (Spenner) Fée

burdock—*Arctium minus* Schk.

bur oak—*Quercus macrocarpa* Michx.

calypso—*Calypso bulbosa* (L.) Oaks

Canadian lynx—*Felis lynx*

caribou, woodland caribou—*Rangifer tarandus*

cedar of Lebanon—*Cedrus libani* A. Rich.

chestnut, American chestnut—*Castanea dentata* (Marsh.) Borkh.

chestnut blight—*Cryphonectria parasitica* (Murr.) Barr

common yellowthroat warbler—*Geothlypis trichas*

cougar, mountain lion—*Felis concolor*

cowbird, brown-headed cowbird—*Molothrus ater*

cucumber magnolia—*Magnolia acuminata* L.

dandelion—*Taraxacum officinale* Weber

deer, white-tailed deer—*Odocoileus virginianus*

Douglas fir—*Pseudotsuga menziesii* (Mirb.) Franco

elk—*Cervus elaphus*

European yew—*Taxus baccata* L.

fireweed—*Epilobium angustifolium* L.

fisher—*Martes pennanti*

flat oatgrass—*Danthonia compressa* Austin

foam-flower—*Tiarella cordifolia* L.

Furbish's lousewort—*Pedicularis furbishiae* S. Wats.

garlic mustard—*Alliaria petiolata* (Bieb.) Cavara and Grande

giant sequoia—*Sequoiadendron giganteum* (Lindl.) Decne.

goutweed—*Aegopodium podagraria* L.

grass-of-Parnassus—*Parnassia parviflora* DC

grey squirrel—*Sciurus carolinensis*

grizzly bear—*Ursus arctos*

grouse, ruffed grouse—*Bonasa umbellus*

gypsy moth—*Lymantria dispar* (L.)

hare, snowshoe hare—*Lepus americanus*

hemlock, eastern hemlock—*Tsuga canadensis* (L.) Carr.

hemlock woolly adelgid—*Adelges tsugae* Annand

hobble-bush—*Viburnum alnifolium* Marshall

hybrid honeysuckle—*Lonicera* X *bella* Zabel

ironwood—*Ostrya virginiana* (Mill.) K. Koch

jack pine—*Pinus banksiana* Lamb.

Karner blue butterfly—*Lycaeides melissa samuelis*

kouprey—*Bos sauveli*

lake trout—*Savelinus namaycush*

loblolly pine—*Pinus taeda* L.

longleaf pine—*Pinus palustris* Miller

lupine—*Lupinus perennis* L.

maples—*Acer* species
mink—*Mustela vison*
Monterey pine—*Pinus radiata* D. Don
moose, forest moose—*Alces alces*
motherwort—*Leonurus cardiaca* L.
mule deer   *Odocoileus hemionus*
muskrat—*Ondatra zibethicus*
Nashville warbler—*Vermivora ruficapilla*
New England violet—*Viola novaeangliae* House
northern goshawk—*Accipiter gentilis*
northern spotted owl—*Strix occidentalis caurina*
oaks—*Quercus* species
paper birch—*Betula papyrifera* Marsh.
parula warbler—*Parula americana*
passenger pigeon—*Ectopistes migratorius*
pileated woodpecker—*Dryocopus pileatus*
pin cherry—*Prunus pensylvanica* L.f.
pine marten—*Martes americana*
pitch pine—*Pinus rigida* Miller
plantains—*Plantago* species
prairie chicken—*Tympanchus cupido*
ponderosa pine—*Pinus ponderosa* Laws.
Port Orford cedar—*Chamaecyparis lawsoniana* (A. Murr.) Parl.
purple loosestrife—*Lythrum salicaria* L.
pseudoryx—*Pseudoryx nghetinhensis*
raccoon—*Procyon lotor*
red-backed vole—*Clethrionomys rutilis*
red-berried elder—*Sambucus racemosa* L.
red cedar—*Juniperus virginiana* L.
red-cockaded woodpecker—*Picoides borealis*
red deer—*Cervus elaphus*
red-headed woodpecker—*Melanerpes erythrocephalus*
red maple—*Acer rubrum* L.
red oak, northern red oak—*Quercus rubra* L.
red pine—*Pinus resinosa* Aiton
red spruce—*Picea rubens* Sarg.
redwood—*Sequoia sempervirens* (D.Don) Endl.

roe deer—*Capreolus capreolus*
round-leaved orchid—*Amerorchis rotundifolia* (Banks) Hultén
royal catchfly—*Silene regia* Sims.
sea lamprey—*Petromyzon marinus*
selfheal—*Prunella vulgaris* L.
sharp-tailed grouse—*Tympanchus phasianellus*
shortleaf pine—*Pinus echinata* Miller
showy lady's-slipper orchid—*Cypripedium reginae* Walter
slash pine—*Pinus elliotii* Engelm.
snail darter—*Percina tanasi*
snipe—*Gallinago gallinago*
solitary vireo—*Vireo solitarius*
spotted knapweed—*Centaurea maculosa* Lam.
spruce budworm—*Choristoneura fumiferana* (Clemens)
sturgeon—*Acipenser fulvescens*
sugar maple—*Acer saccharum* Marsh.
tent caterpillars—*Malacosoma* sp.
thirteen-lined ground squirrel—*Spermophilus tridemlineatus*
thistles—*Cirsium* species
timber wolf, eastern gray wolf—*Canis lupus*
trilliums—*Trillium* species
tuliptree—*Liriodendron tulipifera* L.
walnut—*Juglans nigra* L.
white cedar—*Thuja occidentalis* L.
white pine—*Pinus strobus* L.
whooping crane—*Grus americana*
wild ginseng—*Panax quinquefolium* L.
wild sarsaparilla—*Aralia nudicaulis* L.
wild turkey—*Meleagris gallopavo*
wolverine—*Gulo gulo*
yellow birch—*Betula alleghaniensis* Britton
yeti—(a mythical mammal)
yew, Canada yew—*Taxus canadensis* Marsh.
zebra mussel—*Dreissena polymorpha*

Index